M. ANGRIST 11/92

D0161909

A Short History of Genetics

The Development of
Some of the Main
Lines of Thought:
1864–1939

L. C. Dunn

IOWA STATE UNIVERSITY PRESS ▪ Ames

The late **L. C. Dunn** was Emeritus Professor of Zoology, Columbia University.

Originally published by McGraw-Hill, Inc., © 1965

This edition published by arrangement with McGraw-Hill Publishing Company. New material for this edition © 1991 Iowa State University Press, Ames, Iowa 50010

History of Science and Technology Reprint Series

First printing, 1991

Library of Congress Cataloging-in-Publication Data

Dunn, L. C. (Leslie Clarence)
     A short history of genetics : the development of some of the main lines of thought, 1864–1939 / by L. C. Dunn.
     p.   cm.—(History of science and technology reprint series) Reprint. Originally published: New York : McGraw-Hill, © 1965.
     Includes bibliographical references.
     ISBN 0-8138-0447-7 (alk. paper)
     1. Genetics—History. I. Title.
QH428.D86   1991
575.1'09—dc20   91-18890

Photographic Credits
Brown Brothers: MENDEL—George H. Schull: MENDEL'S PLOT—Genetics: HUGO de VRIES; CARL ERICH CORRENS; ERICH von TSCHERMAK-SEYSENEGG; WILHELM LUDWIG JOHANNSEN; THEODOR BOVERI; WILLIAM BATESON; AR-CHIBALD EDWARD GARROD; WILHELM WEINBERG; EDMUND BEECHER WILSON; T. H. MORGAN and ROLLINS A. EMERSON; CALVIN BLACKMAN BRIDGES—American Museum of Natural History: AUGUST WEISMANN—Victor McKusick: WALTER STANBOROUGH SUTTON.

— Table of Contents —

— Preface —

This book had its genesis in a course of three lectures, The Hideyo Noguchi Lectures, delivered at the Johns Hopkins Hospital, Baltimore, in March, 1964. The purpose of the lectures was to give some idea of the main lines of thought in genetics from the time of Mendel until about 1939. Even with such restriction to genetics in its "classical" period, it was still necessary to be severely selective. Only major principles could be touched on; applications in agriculture and medicine, even normal human genetics could be barely mentioned. Certain important aspects, including cytoplasmic heredity, were left out entirely.

I think this course can be defended on grounds other than lack of time. One of the interesting things about the history of genetics is that a few relatively simple ideas, stated clearly and tested by easily comprehended breeding experiments brought about a fundamental transformation of views about heredity, reproduction, evolution and the structure of living matter. It was chiefly the elucidation of the theory of the gene and its extension to the physical basis of heredity and to the causes of evolutionary changes in populations which gave genetics its unified character. Concentration on these few ideas, especially in brief compass, does not provide the full documented history of genetics which must one day be written. I have been tempted to enter into more detail while preparing this book for publication and have considerably expanded the contents of the lectures; but I have still tried to be brief enough so that the book can be read through as a running account.

The intention was to let the creators of genetics speak for

themselves without extensive discussions of personal histories and peculiarities. I was tempted to break this rule in the case of several of those with strong and interesting characters, but the lack of adequate biographical studies reinforced my decision to restrict personal descriptions and evaluations.

I am grateful to Professor Oswei Temkin, Director of the Institute of. the History of Medicine and chairman of the Committee on the Hideyo Noguchi Lectureship, Professor Victor A. McKusick of the Laboratory of Human Genetics and Professor Bentley Glass of the Department of Biology, all of the Johns Hopkins University for their hospitality and the many kindnesses extended to me during my stay in Baltimore and later during the preparation of this manuscript.

I am grateful to Professor Sewall Wright for generous criticism, from which the manuscript has benefited although he is not responsible for the statements I have made. Professor Walter Landauer has from time to time called my attention to sources of information which I might otherwise not have found, as has Professor John A. Moore. My colleague Professor Dorothea Bennett has given generous help and criticism throughout, and to her and to my wife I am grateful not only for lightening my labors but also for contributions which have given an amateur historian additional pleasure and satisfaction.

L. C. Dunn

New York, August 2, 1965

— Introduction —

THE BIOLOGICAL SCIENCE which has undergone the most rapid development in the first half of the twentieth century and which has most profoundly affected the development of biology as a whole is undoubtedly genetics. It is also a field inadequately treated by historians of science. The need for a general history of genetics has become apparent as its influence on other sciences and on general theories of evolution and development has become more evident, to say nothing of its application to medical, agricultural and sociological problems. The events leading to its rise have been too recent to attract the interest of professional historians, who understandably want the passage of time to aid them in making judgments and in sorting out the ephemeral from the durable.

An adequate perspective is an essential element in all historical research. For those who have participated in the development of genetics, the interest in the unfolding facts and theories and the opportunity to influence its surging progress have in general outweighed any temptation to stand aside long enough to reflect on the origins of its ideas and where they were leading. Such reflections have gone directly into their own research reports, which, together with theoretical discussions, summaries, and textbooks, constitute the raw materials of history-in-the-making. This on the whole is as it should be. A science has to advance before there can be any history of it.

For those working actively within genetics, the accumulating published record of research itself constitutes a history which serves their day-to-day purposes. Part of their stock-in-trade is to know what has been done by their fellows and immediate predecessors. True, as the field diversifies, their knowledge as specialists suffers some constriction; and, as it runs in narrower channels, its relation to the source and to parallel streams of thought is likely to be ignored. Specialists are likely to recognize these tendencies when they are called upon to give instruction in the general field of which their own work forms a part. It is then that they realize the need to overlook the recent and the detailed and to identify the general themes and theoretical framework which give the field cohesion. And while explaining the source of an idea or of a new direction taken by research, they will sometimes be brought up short by the question: But did it really happen in that way?

As the circle of persons interested in genetics widens, the occasions for such reflections multiply. One sees this happening, for example, among biologists having primary interests in other fields who have felt their outlook upon their special problems altered, even threatened, by the strong winds blowing from the perturbed areas of genetics.

Ideas about the mechanism of evolution, a central problem of biology, have been especially sensitive to such disturbances. Was it August Weismann or Gregor Mendel or W. L. Johannsen who made untenable the belief in the inheritance of acquired characters, a belief which solaced most of the biologists of the nineteenth century? What killed orthogenesis, and many other beliefs firmly held by evolutionists without benefit of the concept of the gene? How did the meaning of the word "mutation" change so radically as it passed from paleontology to genetics by way of Hugo de Vries, for whom it signalized the sudden origin of a new species? The idea of mutation threatened to displace Charles Darwin's theory of gradual transformation and thereby alienated from genetics a generation of systematic biologists. And how could reason be restored by assuming that changes in single genes supplied the raw material of evolution when such mutations were usually deleterious?

Once Mendel's theories had become accepted, other biological problems concerned with reproduction, metabolism, development, and the acquisition and maintenance of a specific form had to be reassessed and restated, often with obvious reluctance on the part of those who had reached their views before Mendelism troubled the waters. Animal embryologists were especially disturbed by Mendelism, for their chief problem was to find ways of explaining the integration of a complex set of cells, tissues, and organs into one functioning whole. Mendelism, at first glance, offered only a means of resolving the organism into separable unit characters. Botanists were less restricted by such views since a plant may be viewed as a kind of anarchy anyway. In general, botanists were more receptive to Mendel's ideas; all three of Mendel's "rediscoverers" were botanists. Agricultural research workers were among the first to see their problems in the new light. Mendelism was greeted by many biologists with an interest tinged with opposition.

As research in heredity gathered momentum, scientists in other fields, as well as physicians, scholars, laymen and even politicians, became conscious of the changed aspect which many practical problems assumed when affected by the spread of ideas from genetics. Marked examples of this accompanied the misuse of genetics by the Nazi state under Hitler, and the mistaken denial of some basic scientific truths by political fiat in the USSR.

To all of these sources of interest in the history of genetics may be added one of more general character. The history of genetics provides an example, perhaps the clearest one among the biological sciences, of how scientific knowledge evolves. The logical connection between its ideas—its inner consistency—is readily apparent; and the relation between theory and observation is a matter of open, and usually recent, record. Thus, an account of its growth may serve as a case history for students of the history of science. This has already been done well by Bentley Glass (1963).

It may seem surprising to scientists from other fields that genetics has so infrequently attracted the interest of historians.

Its comparative newness and its specialized language have something to do with this, of course. While these reasons should not deter geneticists themselves, seldom do we find genetics presented in historical perspective. In most textbooks of genetics, ideas are not always, or even often, ascribed to those who first presented or tested them. I can now testify that this creates extra work for those who try to trace such origins. In pursuing such objectives, it has occurred to me that this avoidance of reference to the past, if not deliberate, was at least an attitude of which geneticists were conscious. I once expressed surprise to a fellow geneticist that he seemed unfamiliar with the work of a predecessor. "Well," he said, "if I read everyone else's papers, I wouldn't get my own written." That attitude may be bad for history and it is not a useful view for science generally, but it is at least understandable in a field like genetics, where liberation from restrictions imposed by traditional ideas is sometimes a necessary condition for developing new views. In any case it has had an influence in making genetics always seem like a new or novel science in which history is what has just happened.

This aspect of genetics is especially marked today. The attention of both the scientific and the lay public has for the past ten years been focused on the molecular basis of heredity and on the mode of transmission and transcription of a code of instructions which guides progeny in repeating the biological patterns of their ancestors. The discoveries in this field have been so rapid and exciting and so recent as to create an impression that genetics began in 1944 with O. T. Avery's discovery that the nucleic acid DNA is the vehicle of hereditary transmission. It is possible that Avery's achievement could have been made without benefit of any previous work in genetics. Certainly the discovery by F. Griffith (1928) of genetic transfer from one bacterium to another—which had begun the line of research that led to Avery's experiments—was so made.

Nevertheless the history of the classical period of genetics is of great interest to us, perhaps greater now than ever. Its problems were stated in terms of historical events in organisms—reproduction, heredity, development, evolution—and these problems remain central in biology. We gain deeper understanding of them when they can be stated and studied in molecular terms; but our interest in them will not be exhausted or satisfied by answers at any one level, either molecular or organismal. This means that to the story of the development of earlier ideas, necessarily restricted to the organismal level, there must be added at some future time a chapter on the history of molecular genetics.

There are thus many good reasons, not only scientific and historical, for inquiring into the sources and development of some of the main ideas of genetics. There were also personal reasons that led me in this direction.

I first encountered genetics in 1911, as a college freshman, through reading R. C. Punnett's *Mendelism*. This was fascinating, but it was T. H. Morgan's *Heredity and Sex*, encountered two years later, and discussions of it with Professor John H. Gerould of Dartmouth College, which fixed my interest. The year 1913 was near enough to the beginnings of the modern phase of genetics so that one could easily look back to the time before Mendel's work was rediscovered. The break in the continuity of ideas about heredity around 1900 seemed almost improbably sharp, and that, of course, was a part of the fascination of this new field in which a radical new orderliness seemed to have sprung full-blown from the head of one man.[1] I think I felt some uneasiness about accepting this "one man" view, events in science usually being preceded by some cumulative build-up. There was, in addition, the evidence that each of three "rediscoverers" had independently worked out the essence of the chief principle that Mendel had reached thirty-five years earlier. That made four discoverers even though they were separated temporally.

But such questions, though of recent history, could not be expected to compete, in the minds of young students, with

[1] J. Rostand (1953, p. 242): "On ne connait pas, d'autre exemple d'une science qui soit sortie toute formée du cerveau d'un homme. . . ."

those which were analyzed experimentally in almost every issue of certain biological journals. Something new and exciting was going on. Although I'm not sure that we realized it, a new science was being built. Genetics at the beginning of the second decade of this century was in fact new, its very name but a few years old. Geneticists' interests lay in the future and in the intense activity of the present.

But my misgivings about the suddenness of that beginning remained. They were re-aroused when, as a graduate student in 1915, I became one of the first of a generation to take a degree in the new field. In preparation for it, in seminars at Harvard University presided over by W. E. Castle and E. M. East, I read the works of Darwin, especially the *Variation of Animals and Plants under Domestication* (1868) containing his "Provisional Hypothesis of Pangenesis" and Herbert Spencer's *The Principles of Biology* of 1864, in which he gives his "theory of physiological units." To be sure, the differences between these "units" deduced from speculative principles, and those derived inductively by Mendel from experiments designed for the purpose, were more striking than the similarities. But questions continually recurred as I found that Carl von Naegeli, Weismann, de Vries, and others during the period when Mendel's paper lay unread, had reinvoked and refined these ideas of hereditary units and even associated them with the newly disclosed elements in the nucleus to which the name "chromosome" had just been given. The sharpness of distinction remained, however, in the manner of deriving the ideas of units and the deliberate use of experimental design by Mendel and by those who followed him accentuated the break.

Then I read the works of the nineteenth-century plant hybridizers T. A. Knight, John Goss, Charles Naudin, Karl Friedrich von Gaertner and others who had indeed used experimental methods. When H. F. Roberts published his careful historical essay *Plant Hybridization before Mendel* in 1929, it became apparent how near others had come to the essential solution. They had in fact anticipated certain aspects of it, such as dominance and segregation, but had still not reached a generalized interpretation. To be sure, they helped to define a pathway which seemed more likely to be ancestral to present-day concepts than that taken by speculative philosophers. But how, in that case, was one to regard de Vries, who in the 1880s was both the constructor of a general theory (intracellular pangenesis) and an experimentalist, a plant breeder like Mendel?

Concerning Mendel himself, and the background of his discovery, very little was known. Carl Correns (1922) gave a clear but brief account, "Etwas über Gregor Mendel's Leben und Wirkung," followed shortly (1924) by Hugo Iltis' *Gregor Johann Mendel, Leben, Werk und Wirkung.* The latter was a careful marshaling of the relatively few facts of Mendel's life which could be attested by documents and of reminiscences gathered by Iltis, a botanist and a follower of Mendel, as a teacher in Brünn. It seemed, in 1924, that we were not likely to learn much more about the life of the man who had by that time become a romantic hero, and in fact nothing of importance has since been added to Iltis' account.

Finally, to skip a long period in which my curiosity about these matters was alive but not very active, I encountered through Bentley Glass's review (1957) Alfred Barthelmess' book of 1952, *Vererbungswissenschaft.* After reading the first paragraph in his foreword I concluded that what I had been looking for had been found. That paragraph reads as follows:

This book represents the first attempt to trace the origin and path of development of the science of heredity. Whether one places the date of birth of this branch of biology in the year 1900 or 1866 or even farther back, it nevertheless remains astonishing that until now no history of it has been written. The science of heredity has unfolded itself so precipitately and flowers today so vigorously that one could easily think, in seeking a reason for this lack, that there had been no time for reflection.[2]

2 "Mit dem vorliegenden Band wird zum erstenmal der Versuch gewagt, Ursprung und Weg der Vererbungswissenschaft nachzuzeichnen. Mag man nun die Geburt dieses Zweiges der Biologie auf das Jahr 1900 oder 1866 oder noch weiter zurückdatieren, es bleibt immer erstaunlich, dass bis heute noch keine Geschichte desselben

Those who will read in the footnote the original German of. which the above is a rough translation will see that I have rendered *Vererbungswissenschaft* into English as "science of heredity." Barthelmess himself would translate the term as "genetics."[3] But it seems to me that there should be a distinction between the terms. His first chapter bears the title "Hippocrates und Aristotles die Väter der Vererbungswissenschaft." This indicates to me that all attempts to understand what for centuries was referred to as the "mysterious force" or the "riddle of heredity" are for Barthelmess part of a continuum which culminated in modern genetics. Barthelmess' view requires proof and can hardly be taken as an assumption. To trace the origins of genetics to a time before the beginnings of modern science seems to me to lose sight of its essence. To begin with Aristotle is like beginning a history of some aspect of literature with the origins of the alphabet or of language. Here, I take a narrow view and do not consider the science of genetics as equivalent to thinking about its subject matter—heredity and variation. However, I have no wish to quarrel with the use of terms as such, especially since Barthelmess has provided us with the best guide we have to the history of ideas about heredity. His book may be read with profit by biologists generally and will serve as a valuable source book for historians. Although it has satisfied my own need to a greater extent than any other book, it does not, nor was it intended to, provide a concise account of the ancestry of some of the main ideas developed in classical genetics. A contribution to this end was made by Hans Stubbe in his *Kurze Geschichte der Genetik* (1963). This book is concerned with the period before 1900 and contains an especially helpful chapter on the development of ideas about mutation

---

geschrieben worden ist. Die Vererbungswissenschaft hat sich so stürmisch entfaltet und blüht gerade heute so reichlich, dass man, nach einem Grund dieses Fehlens forschend, leicht dem Gedanken Raum geben möchte, sie hätte noch gar keine Zeit zur Besinnung gehabt."
[3] He says (p. 224): ". . . die Genetik—dieser kurze und praktische Ausdruck Batesons hat sich inzwischen für die Vererbungswissenschaft international eingebürgert. . . ."

in the nineteenth century. Since Barthelmess and Stubbe have given us materials for a history of some of the problems encountered in thinking about heredity, the preparation of this account is made much easier, and briefer treatment for a wider audience is made possible.

Since I have now given some indication of the kinds of motives which set me off on this task, I should now describe in somewhat more detail how I propose to proceed.

I think what interests me most in the history of science is the relationship between ideas held at different times, couched in similar terms, yet obviously having different contents and meanings. The view of matter as composed of elementary corpuscles, atoms, preceded by two millennia the development of atomic theory. What, if anything, does the second concept owe to the first? How, if not derived from the first, did the second arise?

The most important concept in genetics is the view that the organization and activities of living matter rest on a system of self-replicating living units. Living units ("physiological units," "gemmules," "micellae," "pangenes") appeared in biological thinking long before Johannsen introduced the term "gene" in 1909.

The earlier forms of the concept, in physics as in biology, were vague and unfruitful in the sense that today's theories could not be derived directly from the earlier ideas. It is precisely for this reason that an inquiry into the development of these ideas is useful and interesting to us today. To question the continuity of the content of ideas phrased in similar words puts the emphasis where it belongs. It is the analysis of the operative content rather than the form of statement which approaches the essence of modern science. To trace the continuity, the pedigrees, of ideas is only to set the stage, a necessary and often interesting operation, but one which has only formal or academic value if it does not lead to some disclosure of what made the ideas evolve in the direction of the forms in which we see them today.

One reasonable ambition would be to put a succession of

historical events in outline form. But risk attends this operation: the events may seem related because successive and the
facts of historical sequence may suggest to the reader notions
of how earlier events determined later ones. What is even
worse is the possible effect such a list could have on the mind
of him who draws it up. Hypotheses are useful and necessary
for the scientist; but for the historian, whose first concern
should be the accuracy and completeness of the record, they
are fraught with the danger of digression. The leaps of imagination by which the gaps between events are bridged have
to be seen for what they are—short-cuts through the air. They
must be subject to the strict restraint which seasoned historians call "discipline."

In tracing historical sequences, then, the desire to identify
causal connections relating later to earlier forms of ideas can
hardly be suppressed, but I for one can hardly expect it to
be gratified very often. One has as raw material only what
men wrote and published at particular times. How much the
different contributors knew about the work of predecessors
or contemporaries can usually only be determined by what
they said they knew, and by inferences, usually subject to
uncertainty, from the state of general knowledge and the
intellectual atmosphere of the times.

For example, in tracing the development of the idea of
living units, which underwent so sudden a transformation beginning with Mendel, how are we to know what influence was
exerted upon him by those who taught him physics and
chemistry, in which movement was already conceived in terms
of units? Or again, there appeared in his symbolic notation
of the distribution of differentiating characters among the
offspring of hybrids what seems to be the first described case
of the binomial theorem operating in living systems. Shall we
list that anciently known relationship among the "causes" of
the transformation? On the whole, the most profitable course,
at this stage, would seem to be to list the successive forms in
which the main idea appeared and to compare their contents
in relation to the form which the idea has now assumed in

the developed theory. Thus, we should proceed in the usual manner of scientific inquiry, asking *how* things happened rather than *why*.

I should not suppose that in genetics there are very many key concepts to which this sort of historical inquiry can be applied. This statement probably reveals my opinion that, in the sense in which I have been using the term, there are not very many "ideas" in genetics altogether, which in turn means that I distinguish general concepts from hypotheses, dozens of which are conceived each year.

What then are the ideas with which I propose to deal? First, of course, there is the concept of replicating living units as the basis of living activity. This, for the period which I shall cover, may be stated more succinctly as the *theory of the gene*. Second, there is a view of the pattern or system in which the genes are arranged in the transmission mechanism. Third, there is the view that change in the properties of a replicating element occurs in a certain way, by mutation. Fourth, there are ideas connecting changes in the arrangement and properties of the elements with evolutionary changes in species and other populations. And fifth, there is the view that genes control metabolism and development by modulating stepwise sequences of processes. These ideas correspond to the major problems with which speculation and, later, investigations of heredity and variation have traditionally proposed to deal: perpetuation or maintenance of common properties in descent; alteration and diversification of these among populations distributed in space and time; and the problem of the regulation or control of growth in the individual.

These ideas obviously should be looked at from a point in time at which they have assumed a definite form, although not necessarily a final one. The point from which I have chosen to look back is 1939, chiefly because by this time the main ideas could be recognized. We may say that in the preceding period back to about 1900, genetics as we know it today took a form sufficiently unified to constitute a distinctive entity within biology. I shall call this the period of "classical genetics." The linear arrangement of genes had been proved, first as a self-consistent, logical interpretation of the results of breeding experiments and, second, as a property of the chromosome mechanism revealed by the rapid development of cytology. Confidence in this view was established in the middle 1930s when, largely as a result of refinements in cytological theory and technique, including analysis of the giant chromosomes of the salivary glands of Drosophila and of the cytogenetics of maize, the linkage maps were shown to correspond to visible chromosomes. By 1939, a method for the study of the mutation process had been discovered and tested by quantitative studies of mutations induced by radiations; the mathematical theory underlying the behavior of the genetic system in populations had been definitively outlined and reconciled with the theory of natural selection so that evolution could be interpreted as an outcome of processes open to testing by observation and experiment; the first gene-controlled synthetic sequences in development had been proved and the foundation laid for extensions of the theory of the gene into cellular physiology, biochemistry, metabolism and growth.

I shall not at this point specify individual publications marking the stage of attainment in each of the subjects mentioned above. In fact, the late 1930s was a period of new beginnings rather than of culminations or endings, and that was its importance. Certain events and summary publications indicating this character of the period will be discussed later. For present purposes in defending the choice of 1939 as a cut-off point, I should emphasize that by this time the theory of the gene had taken on a more general character. The gene had become an essential term in describing continuity, function, evolution. This expansion in the range of phenomena encompassed by the idea, much greater than could have been foreseen in 1900, had been accomplished by about 1939 without losing the fundamental unity implicit in the original concept of the gene. The fruitfulness of the general theory had been proved by the best of tests, that is, its predictive value when applied to relations not yet tested by observation or experiment, such as the behavior of parts of chromosomes when broken by radiation, and the success achieved when it was applied to explain observations made long before the theory was developed. A conspicuous case of the latter sort was the clearing up of the puzzling behavior shown by the variant forms of Oenothera species which de Vries had begun to describe in 1887.

Setting an end date, such as 1939, for the survey of any segment of scientific thought is a simple matter compared to the problem of choosing where to begin. It is obvious enough that written history is open at both ends—while one may cut it off even in an arbitrary way, to make judgments about beginnings must await the results of research.

What we seek to discover about the origins of genetics will depend of course on the kind of definition of that field which guides us. I have been guided by the belief that a chief use of tracing the history of a set of ideas like those embodied in genetics is to lead us to a better understanding of genetics as it is today. That purpose, it seems to me, is best served by applying as a criterion for selection the survival of ideas about heredity which form essential parts of current theory. One is thereby aided by a kind of natural selection which tends to dispose of less profitable ideas. This calls for a sharper focus, a narrowed definition of genetics, and a limited time span.

Genetics, dealing with the main ideas I have mentioned above, came into existence only after the rediscovery of Mendel's principles in 1900. The name "genetics" was first used by William Bateson in 1906 (cf. p. 69), but the study of the units which Mendel had discovered was, in the period before about 1915, usually referred to as "Mendelism." Later, genetics itself was often qualified as "Mendelian genetics." Now it can be understood to include all studies of heredity and variation directed toward elucidation of the transmission mechanism of heredity (formal genetics), the functioning of

hereditary elements in development and metabolism (physiological genetics), and the dynamics of distribution and change
of such elements in populations and in evolution (population
genetics).

I shall begin my account with certain events in the spring
of 1900 and with the state of knowledge of heredity and
variation at that time. I will then move back some 40 years
to trace the earlier states of the concepts which after 1900
made possible that extraordinary burst of experimental activity which by 1920 had solved the problem of the transmission mechanism of heredity. The period of fruition—roughly
the forty years following 1900—will then be reviewed. The
whole period of the "history" is seventy or eighty years, but
only some thirty-five of these will be dealt with in any detail.

Years, however, are poor measures of duration in the history of science, just as linear distance fails to describe the
amount of effort required to go from one place to another
in mountainous country. Dates are reference points, useful
but not enlightening. Sequence and order are more important.

If the time-depth is shallow, so, too, in the restricted study
I have in mind will be the degree of philosophical penetra-
tion. The results achieved in the development of genetics
were derived from the application of experimental methods,
but the question which experiment will not answer is what it
all means. The hope for an answer to such a question, whether
expressed or lurking in the background, may be an actual
handicap to investigation. We may find ourselves unready or
unwilling to accept the limited, either-or kind of decision
which experiment is competent to provide and thus lose the
solid substance of discovery for the shadow of a larger hope.
It is not that I, or another who may review the development
of the science of heredity, may not be interested in what it
means, but rather that before we ask that ultimate question we
should know what has caused it to be asked.

— Preface —

TO THE REPRINT

IN THE EARLY 1960s, Alfred H. Sturtevant and Leslie C. Dunn, two of the outstanding geneticists of their generation, were racing to see who would be the first to get his history of genetics into print. Dunn's *Short History of Genetics* (the longer work) appeared in 1964, and Sturtevant's *A History of Genetics* (Harper and Row) in 1965. The rivalry was friendly and in any case the result was two complementary books.

Dunn had been a graduate student in the Edward Murray East–William Ernest Castle group at the Bussey Institution of Harvard University, and Sturtevant was in the Thomas Hunt Morgan group at Columbia University. These two groups were chief rivals in the field of genetics during the period 1910–1930. Dunn and Sturtevant each wrote from his own scientific perspective. Of the two books, Dunn's is clearly the broader in historical perspective.

Dunn was unhappy with his book from the moment it was published. He knew that the history of genetics in the twentieth century was very complex and that his *Short History*, despite being a substantive start, was hardly complete. In June 1973, Dunn embarked upon a major revision of his book. He was eagerly working on this project when he died suddenly in the spring of 1974, about a month before the first part of Ernst Mayr's conference on evolutionary synthesis, to which Dunn was invited.

It is unfortunate that Dunn was unable to complete his revision; new sections focused on the period 1939–1960 and would have been especially valuable. But in its original form, Dunn's *Short History* remains the single volume of greatest help in understanding the history of genetics in the period 1900–1939.

For the reader who wishes to extend his/her understanding of genetics in the twentieth century, I suggest consulting the following books, all of which have extensive bibliographies or bibliographic essays: Garland E. Allen, *Life Science in the Twentieth Century* (Cambridge University Press, 1975) and his *Thomas Hunt Morgan: The*

Man and His Science (Princeton University Press, 1978); Elof Axel Carlson, *Genes, Radiation, and Society: The Life and Work of H. J. Muller* (Cornell University Press, 1981); William B. Provine, *The Origins of Theoretical Population Genetics* (University of Chicago Press, 1971) and *Sewall Wright and Evolutionary Biology* (University of Chicago Press, 1986); and Jan Sapp, *Beyond the Gene: Cytoplasmic Inheritance and the Struggle for Authority in Genetics* (Oxford University Press, 1987).

In addition to his *Short History,* Dunn left an extensive oral history memoir (Columbia Oral History Program) and was very active in saving the papers of geneticists (of course including his own papers) at the Library of the American Philosophical Society in Philadephia, where the collection on the history of genetics is far and away the best in the world.

William B. Provine
PROFESSOR OF ECOLOGY AND SYSTEMATICS
AND IN THE DEPARTMENT OF HISTORY
CORNELL UNIVERSITY

Part I

THE BIRTH OF GENETICS

– Chapter 1 –

MENDEL AND HIS DISCOVERY

The Rediscovery of Mendel's Principles

ON MARCH 14, 1900, the editor of the *Berichte der deutschen botanischen Gesellschaft (Reports of the German Botanical Society)*, received a manuscript with the title "Das Spaltungsgesetz der Bastarde" ("The Law of Splitting of Hybrids"), by Hugo de Vries (1848–1935), professor of botany in the University of Amsterdam. It was submitted for presentation at the March 30 meeting of the Society and was printed in the issue of the *Berichte* which appeared on April 25. After submitting this manuscript as a "preliminary report," de Vries sent a shorter version, a summary of the German paper, in French, to the Paris Academy of Sciences. This appeared in print in the *Comptes Rendues de l'Academie des Sciences* on March 26, that is, before the German paper had appeared.

This summary French paper is usually cited as the first announcement of the rediscovery of Mendel's principles. In fact, Mendel was not mentioned in it, although de Vries had acknowledged in the German paper that he had learned of Mendel's paper of 1866, but only after the greater part of his own researches had been completed and the conclusions derived.

It was de Vries' French paper also which started the chain

reaction that brought the work of Carl Correns and of Erich von Tschermak-Seysenegg into print in that same spring of 1900. Correns (1864–1933) was then dozent in botany at the University of Tübingen, where since 1894 he had used the methods of experimental breeding to study problems of heredity in maize, snapdragons, beans, peas, and lilies. By 1899, he had solved the problem of xenia (direct influence of foreign pollen on the character of the seed it fertilizes) and in so doing had carried out a genetic analysis of races of maize and peas. The explanation of the splitting, or segregation, of characters in the offspring of hybrids, which he had observed in maize and in peas, came to him, he tells us, in a sleepless night of November, 1899. Searching through the literature the next day, he found in W. O. Focke's *Pflanzenmischlinge* (1881) a reference to Mendel's paper and realized that his "explanation" had been anticipated thirty-four years before. He thus felt there was no urgency in publishing his own results, particularly since he had found evidence of absence of dominance. In this respect his results differed from Mendel's, and he thought therefore that further work on his part was needed.

However, on April 21, 1900, Correns received from de Vries a copy of his report to the Paris Academy, and, on the 22nd, he sent a short report of his own work and main conclusions to the German Botanical Society with the title "G. Mendel's Regel über das Verhalten der Nachkommenschaft der Rassenbastarde" ("G. Mendel's Law Concerning the Behavior of Progeny of Varietal Hybrids"). It appeared in the *Berichte* in the third week of May.

On June 2, 1900, the editor of the *Berichte* received a manuscript entitled "Über künstliche Kreuzung bei Pisum sativum" ("Concerning Artificial Crossing in *Pisum sativum*"). It had been sent in for publication by Erich von Tschermak-Seysenegg, a volunteer assistant at the Austrian Imperial Family Foundation Estate (a kind of agricultural experiment station) at Esslingen, near Vienna. It appeared in the July issue of the *Berichte*. Tschermak, however, had already written up his results as an inaugural dissertation submitted to the Hochschule für Bodenkunde (Agricultural

College) on January 17, 1900. When de Vries' first report appeared, Tschermak urged the immediate printing of his dissertation, and it appeared in May, 1900, in the *Zeitschrift für das landwirtschaftliche Versuchswesen (Journal of the Agricultural Research Establishment)* even before the shorter version in the German *Berichte*, which is usually cited as the third of the "rediscovery reports."

Tschermak, the youngest of the three rediscoverers (born 1871), had begun experimental breeding research only in 1898 in the Botanical Garden of Ghent, in Belgium, in an effort to test some of Darwin's ideas on the effects of self- and cross-fertilization. In 1899, he found the law of segregation in F_2 populations—in the second generation of offspring—and only afterwards found Mendel's paper. "Even in the second year of experimentation," he wrote in a postscript to his first paper, after having seen Correns' and de Vries' publications, "I too still believed that I had found something new."

Thus, within the space of two months (late March to late May), the first laws of heredity established by experimental methods leaped from the obscurity of three decades' neglect. Of primary importance was the fact that they had been confirmed by the unusual process of rediscovery not once but three times and independently.[4]

Mendel's Discovery

What in fact had Mendel discovered? Our primary source is Mendel's paper of 1866, a concise transcript of two reports

[4] The best source in English for the texts of the papers of de Vries, Correns, and Tschermak, as well as for the correspondence of Mendel and Naegeli (see below), is *Genetics*, Volume 35, 1950, supplement to Number 5, Part 2, pp. 1–47, printed as a separate section in connection with the celebration of the fiftieth anniversary of the birth of genetics.

The circumstances of the rediscovery are described and letters from de Vries, Correns, and Tschermak are printed in H. F. Roberts (1929). The most recent review of pre-1900 genetics (Hans Stubbe, 1963) contains a helpful and judicious chapter (IX) on "Die Wiederentdeckung u.s.w." The bibliography in the last cited work is fairly complete, but additional information on Correns is given by Emmy Stein (1950).

he had given at the Natural Science Society of Brünn on February 8 and March 8, 1865. The whole story of the development of the new theory is given clearly in these forty-four printed pages. No better account of his work has ever been written, and whoever wishes to understand the essential nature of genetics should begin by reading it. There is also a letter of Mendel's (*cf.* Correns, 1905) to the Swiss botanist, Carl von Naegeli, dated April 18, 1867, in which Mendel attempts to explain the meaning of his paper to the leader of European botany. Mendel had sent Naegeli a reprint of his paper, and it is obvious from the letter that Naegeli had not appreciated the import of Mendel's interpretation of his results.

Mendel's paper illustrates the crucial importance, in a scientific inquiry, of the proper framing of the question at issue, what German writers call the "Fragestellung." In Mendel's case this was unambiguously stated. The object of his investigation, stimulated by the regularity with which similar hybrids appeared from crosses between the same species, was to follow up the development of the progeny of the hybrids

> . . . in such a way as to make it possible to determine the number of different forms under which the offspring of hybrids appear—to arrange these forms with certainty according to their separate generations—definitely to ascertain their statistical relations.

A primary condition for discovering a set of statistical rules was to have set out to discover them. Such questions had, as Mendel points out, never been asked before.

Mendel then tells us why the pure-breeding varieties of peas ("pure lines," we should now say) provided experimental material meeting the requirements imposed by the questions. Thirty-four varieties were tested for two generations for constancy of certain "differentiating characters" under self-fertilization. These characters were arranged in contrasted pairs such as round *vs.* wrinkled seeds, tall *vs.* dwarf stature, and so forth. Varieties differing in one pair of these characters were crossed and the hybrid offspring classified and allowed to self-fertilize to produce second and third generations. The plants thus produced were classified according to the contrasted pairs of traits and the numbers in each class were counted.

The results, now familiar from their frequent repetition in biology textbooks, showed a remarkable orderliness. The plants derived from the first cross were uniform in respect to the character studied and in the cases of seven contrasted pairs all resembled, in this character, one of the parents. The character appearing in this first, or F_1, generation of offspring was called "dominant." One other pair of characters, late as contrasted with early flowering habit, was not reported in detail since the results were incomplete. In this case, the hybrid was intermediate between the parents, while the off-spring of the hybrids "probably follows the rule ascertained in the case of other characters." Mendel apparently realized that dominance was neither invariable nor related to the method of inheritance.

The hybrids when inbred (self-fertilized) produced an F_2 generation consisting of two phenotypes, $3/4$ with the dominant and $1/4$ with the recessive character of the pair. But these were shown by their (F_3) offspring to consist of three genetically different classes: $1/4$ true-breeding dominants, $2/4$ dominants which did not breed true but again produced offspring splitting into $3/4$ dominants and $1/4$ recessives, and $1/4$ true-breeding recessives. It was Mendel's establishment of this $1:2:1$ ratio which constituted the essential feature of his discovery:

it is now clear that the hybrids form seeds having *one or the other* of the two differentiating characters and of these one-half develop again the hybrid form while the other half yields plants which remain constant and receive the dominant or the recessive characters (respectively) in equal numbers.

The above is cited from the English translation of Mendel's paper made by the Royal Horticultural Society and printed in its Journal (1901, Vol. 26, pp. 1–32) with footnotes by

William Bateson. This was the first appearance of Mendel's paper in English and it has been frequently reprinted in British and American textbooks. Dr. Alan Robertson of the Department of Genetics, University of Edinburgh, has called my attention to certain inaccuracies in this translation and especially to one in the section cited above. The original German is: "wird es nun ersichtlich, dass die Hybriden je zweier differierender Merkmale Samen bilden, von denen die eine Hälfte wieder die Hybridform entwickelt, während die andere Pflanzen gibt, welche konstant bleiben und zu gleichen Teilen den dominierenden und rezessiven Charakter erhalten."

The German clearly emphasizes the descriptive, phenotypic character of the observations, which refer to zygotes (seeds): "the hybrids of each two differentiating characters form seeds of which the one half again develops the hybrid form, while the other yields plants which remain constant and in equal portions receive the dominant and recessive character," whereas the English words "one or the other," which do not appear in the German, suggest the segregation mechanism in the formation of gametes which is not observation but interpretation.

The essential theory is stated later in the section on the reproductive cells of the hybrids (pp. 24–25 of the original) "dass die verschiedenen Arten von Keim und Pollenzellen an der Hybride durchschnittlich in gleicher Anzahl gebildet werden" ("that the different kinds of egg and pollen cells are formed in the hybrid on the average in equal numbers").

The statement of observation—equal proportions of differentiating *characters*—is thus translated into theory as "equal numbers" of *gametes* bearing the alternative traits. The existence of alternatives such as tall and short had been described in advance of the crossing experiments. The experimental tests showed that the seven pairs reported had been correctly paired, and that the essential mechanism was concerned with the formation of egg and pollen and the partitioning or splitting or, as we now say, "segregation" of the alternatives to different cells.

In order to make this essential point clear, Mendel repeated it in several symbolic forms. In respect to one difference, A (dominant) versus a (recessive), the occurrence in the F_2 generation of three forms in the ratio $\frac{1}{4}AA:\frac{1}{2}Aa:\frac{1}{4}aa$[5] means that for each four fertilizations, the constitution of the egg and pollen cells and the outcome of these should be as follows:

pollen cells	A	A	a	a
egg cells	A	a	A	a
offspring	AA	$2Aa$		aa

The offspring ratio is thus "explained" by random union, in fertilization, of gametes half of which are pure for one and half for the other of each pair of alternatives, such as A or a.

Mendel repeated all this to Naegeli, pointing out that the differences among the germ cells of the hybrids, which he regarded as the essential point, were based on experiments and not derived *ad hoc*.

He similarly derived from experiments the rule concerning inheritance when, in hybrids, more than one pair of alternatives was observed. Here each pair behaved in inheritance independently of the other, F_2 consisting of $\frac{3}{4}$ with the dominant member and $\frac{1}{4}$ with the recessive of each pair taken separately. The combinations appeared in the proportions: $\frac{9}{16}$ with both dominants (AB), $\frac{3}{16}$ with one (Ab), $\frac{3}{16}$ with the other (aB), and $\frac{1}{16}$ with both recessives (ab). Mendel explained this on the basis of the composition of the gametes by supposing all combinations, AB, Ab, aB and ab, of both egg and pollen cells to be formed in equal numbers and to unite in fertilization at random. He showed that the F_2 classes could be derived by multiplying the two single-character F_2 ratios together:

$$AA + 2Aa + aa$$
$$BB + 2Bb + bb$$

[5] Mendel used only single letters A or a to represent the "constant" forms which we now symbolize unambiguously as AA and aa.

Producing: 1*AABB*   1*AAbb*   1*aaBB*   1*aabb*
            2*AaBB*   2*Aabb*   2*aaBb*
            4*AaBb*
            2*AABb*
            _____
            9*A*()*B*() 3*A*()*bb* 3*aaB*() 1*aabb*

Then he proved the equal frequency of the four kinds of gametes more directly by crossing hybrids *AaBb* with recessives *aabb* and showing that the offspring of the four combinations *AB*, *Ab*, *aB*, and *ab* occurred in equal numbers.

Finally, he studied the inheritance of three pairs of alternative traits in one experiment by crossing a variety having *round* seeds with *yellow* cotyledons and *grey* seed coats with another variety having *wrinkled* seeds with *green* cotyledons and *white* seed coats. These appeared in the offspring of the hybrids in the proportions:

27/64	with	three	dominant	characters, e.g.,	*ABC*
9/64	with	two	dominant	characters, e.g.,	*ABc*
9/64	"	"	"	" "	*AbC*
9/64	"	"	"	" "	*aBC*
3/64	"	one	"	" "	*Abc*
3/64	"	"	"	" "	*aBc*
3/64	"	"	"	" "	*abC*
1/64	"	no	"	" "	*abc*

Mendel showed that this could be derived by multiplying the three series:

        1*AA*   2*Aa*   1*aa*
        1*BB*   2*Bb*   1*bb*
        1*CC*   2*Cc*   1*cc*

The rule was that in hybrids which form egg and pollen cells with contrasted characters in equal numbers, all possible combinations of members of different pairs are formed at random.

The two parts of "the law discovered for Pisum" as Mendel referred to it, were thus the splitting, or segregation, of alternative members of each pair of differentiating characters into different gametes, and the independence in this process of members of different pairs, now usually referred to as "independent assortment." The first was obviously a prior condition for the second and is now generally recognized as the basic and most general principle of heredity. It should be noted, however, that proof of the existence of a transmissible element as a separable entity depends upon proof of its actual separation. Recombination of elements is thus an essential feature of the theory.

For proof of these rules Mendel relied not merely on showing that each pair of contrasted characters appeared in the offspring of a hybrid (the F_2 generation) in a ratio of three dominants to one recessive. He tested samples of F_2 plants showing the dominant character of each pair and showed that ⅓ of them gave only offspring with the dominant character while ⅔ again segregated as the F_1 had. The first were thus homozygous (AA) and the second, heterozygous (Aa) in a ratio of 1:2 as predicted by his theory. He analyzed in a similar way the members of the F_2 generation from the dihybrid ($AABB \times aabb$) and trihybrid ($AABBCC \times aabbcc$) crosses and found that the results fitted what the theory predicted very closely indeed.

This could be taken as proof that three pairs of differentiating seed characters, namely round vs. wrinkled seed form, yellow vs. green cotyledon color, and grey-brown vs. white seed-coat color were inherited independently of each other, that is, recombined at random when the germ cells of the trihybrid were formed. Such statistical proof was given for independent inheritance of only three of the seven pairs of characters. Of the other four pairs of characters Mendel reported, without giving numerical data: "Further experiments were made with a smaller number of experimental plants in which the remaining characters by twos and threes were united as hybrids: all yielded approximately the same results." His assumption that all seven pairs were inherited independently was thus not strictly justified. He saw clearly enough what the principle of independent assortment meant, namely that characters "may be obtained in all the associations which are possible according to the [mathematical] laws of combination." In the case of the seven characters he had studied in peas, he estimated the number of combinations as 2^7, or 128, and indicated that he had obtained them all.

Since it has turned out that there are seven pairs of chromosomes and seven groups of linked genes in the pea plant, the chance that Mendel would have chosen one gene pair from each of the seven groups would be $9/7 \times 5/7 \times \ldots \times 2/7 \times 1/7$, or about $1/163$. This is sometimes taken as an estimate of "Mendel's luck": the chance that Mendel would have encountered just such an unlikely distribution, and thereby made the discovery of linkage which was reserved for Bateson and Punnett in 1906.

In 1936, R. A. Fisher made a statistical study of Mendel's published results and found that something more than luck must have been involved since the "goodness of fit" of data to theory was almost impossibly good (total χ^2 having p = .99993). In the tests of 1:2 ratios from F_2 dominants, the fit was close to the ratio which Mendel expected, but not to the ratio which should have been expected. Since only ten seeds were grown from each plant, in some five or six percent of the cases the plant could have been Aa and yet have produced, by chance, ten offspring with the dominant trait and thus would have been classed as AA. But in all such cases the results fit Mendel's uncorrected expectation.

Several explanations can be suggested. First, that Mendel had already reached his theory before the published tests were made and knew what to expect. In the later experiments which fit the theory almost perfectly, he may have stopped counting when the expected result was obtained. Second, he may, on the basis of prior experience, have chosen seven pairs of contrasted characters that had shown independent inheritance. Third, in regard to the great discrepancy in the 1:2 ratios, in Fisher's view, "it remains a possibility that Mendel was deceived by some assistant who knew too well what was expected." There is no evidence for this, but no other simple explanation of the discrepancy is available. Opportunity for some unintentional selection may have existed, for in describing the observations on length of stem Mendel says, "In this experiment the dwarfed plants were carefully lifted and transferred to a special bed" to prevent them from being overgrown by their tall relatives.

The impression that one gets from Mendel's paper itself and from Fisher's study of it is that Mendel had the theory in mind when he made the experiments reported in the paper. He may even have deduced the rules from a particulate view of heredity which he had reached before beginning work with peas. If so, the outcome of his experiments constitutes, in Fisher's words, not discovery but demonstration.

It is clear that Mendel had great confidence in his results since he had tested and confirmed his theory in all ways that occurred to him, using different races of peas. His tests of the theory by crosses among varieties of beans in general agreed with those with peas, although he found one character in beans, affecting flower and seed color, which gave results at variance with the simple monofactorial explanation derived for a similar difference in peas. Nevertheless, Mendel had sufficient confidence in the general principle to suggest that since the recessive color (white) appeared in much less than a quarter of the F_2 generation, perhaps it depended not on one but on two independent factors which would be expected in combination in only $\frac{1}{16}$ of the offspring of the hybrids. This first example of an apparent exception due to the interaction of different factors on one character was confirmed forty years later, after the "rediscovery." Mendel also discussed the manner in which the principles discovered in peas might apply to other plant hybrids and concluded:

Whether the variable hybrids of other plant species observe an entire agreement must also be first decided experimentally. In the meantime we may assume that in material points an essential difference can scarcely occur, since the unity in the developmental plan of organic life is beyond question.

From Mendel's own published work we should never have known the extent of the efforts which he devoted in the years following 1865 to further testing of his theory. He published only two botanical papers, the famous one of 1866, and a short one in 1869 reporting on the ill-fated experiments with the hawk-weed species (Hieracium), which failed to confirm, and, as we know now, even to test, his theory. Our information about Mendel's work from 1865–1873 is derived from his ten letters to Naegeli, written between 1866 and 1873. These letters were obtained from Naegeli's family and published with a short supplement by Correns (1905). They were evidently written as scientific papers. They report hybridization experiments between species or varieties belonging to twenty-six different genera. In some cases, the rules derived from Pisum were clearly confirmed. In others, however, they were not. The most notable of these exceptions was provided by the genus to which Mendel had devoted his chief efforts for six years, that of the hawk-weeds, Hieracium. Mendel's unfortunate choice of this intractable material was encouraged by Naegeli, one of the chief authorities on this group of plants, who urged Mendel to concentrate on it and sent him seeds and plants. After five years of intensive work which had diverted him from other studies and ruined his eyesight, he wrote to Naegeli (July 3, 1870) : "On this occasion I cannot resist remarking how striking it is that the hybrids of Hieracium show a behavior exactly opposite to those of Pisum."

This modest understatement must have concealed a disappointment that foreshadowed the end of Mendel's scientific work. It was not made clear until forty-five years later, by two Swedish investigators, that species of Hieracium produce offspring in part by apogamy, that is, by the botanical near-equivalent of parthenogenesis, and in part by normal fertilization, so that crossbred offspring can be, but are not always, formed. What defeated Mendel, however, was the fact that in *hybrids* of different species of this genus, the flowers are *always* apogamous, that is, the parent reproduces vegetatively and the offspring are all alike, as though derived from cuttings, and no sexual process, and hence no segregation, can occur. And this was the genus which Mendel, on the advice of Naegeli, had chosen to provide the chief test of his theory! No wonder that another experimental geneticist, nearly 100 years later should say "The encounter with Naegeli became a disaster for Mendel" (Renner, 1959).

In the midst of this work of testing his theory, Mendel became the spiritual and administrative head of his monastery. On May 4, 1868, he wrote to Naegeli, "Recently there has been a quite unexpected change in my circumstances. On March 30, my humble self [meine Wenigkeit] was elected by the chapter of the foundation to which I belong to be its head for life." Nevertheless, he continued his exacting hybridization work for another three years, kept up his meteorological observations, and in 1870 delivered his last paper—on the Brünn tornado of 1870—at the Natural Science Society of Brünn and published it in the *Proceedings* in the following year.

The last years of Mendel's life were spent in a futile battle with the Austrian government over the right of the state to tax the monastery for support of public worship. Mendel, a liberal if not a radical, refused to pay, and fought the ministry until his death, after which a compromise was worked out. His embroilment in administration and in this matter of conscience caused the termination of his scientific work. Until his death, in 1884, none of his work, either scientific or administrative, seemed to him to have borne fruit, although his brethren of the cloister and his fellow townsmen expressed their love and respect for a man of integrity and courage.

After the publication of the first paper of 1866, there followed one of the strangest silences in the history of biology. The paper was cited in two reviews: in Hermann Hoffmann's "Untersuchungen zur Bestimmung des Wertes von Species und Varietät" (1869), and in Focke's *Pflanzenmischlinge* (1881), but without any indication that these authors had grasped the import of Mendel's work. Two further citations before 1900 were in the bibliography of the article "Hybridism," by G. J. Romanes, in the 9th edition (1881–1895) of the *Encyclopedia Britannica*, Volume 12, pp. 422–426, and in the bibliography of an article "Crossbreeding and Hybridization" by L. H. Bailey, published in 1892. Bailey reported that he had taken the title of Mendel's paper from Focke and had never seen the paper itself. De Vries found the reference to Mendel in Bailey's bibliography (*cf.* Roberts, 1929, p. 323).

The Proceedings of the Brünn Natural Science Society were sent to 120 learned societies, academies and libraries in the Old World and the New. In 1928, on the occasion of depositing in the Columbia University library a copy of the volume for 1866 (Volume 4) containing Mendel's paper, which I had obtained from a bookseller in Germany, I was surprised to find that a set of the Brünn proceedings had been there at least since 1897, but Volume 4 had apparently never been consulted. It seems to have met a similar fate in most of the twenty-one other American libraries in which Volume 4 has been found (Dorsey, 1944).

On Mendel's manuscript, the editor of the *Proceedings* had penciled the annotation "40 reprints." One of these we know went to Carl von Naegeli and another to Anton Kerner von Marilaun, who after Naegeli was the chief authority of his period on hybridization. Tschermak, however, has reported (Stubbe, 1963, p. 110) that Kerner's assistant found Mendel's reprint in Kerner's library with the pages uncut. Another copy was sent to de Vries in 1900 by M. W. Beyerinck, the professor of bacteriology at the University of Delft (Stomps, 1954), who had himself suspected the operation of something like unitary mutation in bacteria. It is apparent that neither the journal in which it appeared nor Mendel's paper itself were "unknown" before 1900. "Unappreciated" is the more correct description of its fate.

The circumstance which most commentators have considered to be the strangest of all is that Naegeli, one of the foremost botanists and hybridizers of his time, failed to mention Mendel or his theory either in papers published after the extensive correspondence with Mendel or in his major work on heredity (Naegeli, 1884). Although Naegeli's letters to Mendel have not been found, his objections to Mendel's theory were quoted by Mendel in his reply to Naegeli (letter of April 18, 1867) :

1. Whether one may conclude that constancy of type has been obtained if the hybrid Aa produces plant A and this plant in turn produces only A.

2. You should regard the numerical expressions as being only empirical, because they cannot be proved rational.

These and other statements of Naegeli's, coupled with the absence of comment on Mendel's theory in Naegeli's subsequent works, gave rise to the view now commonly held that Naegeli had missed the essential feature of Mendel's demonstration and did not recognize the significance of Mendel's work.

Alexander Weinstein (1958, 1959, 1962) has examined the contemporary documents bearing on this question and has given reasons for his opinion that Naegeli had in fact understood Mendel's discovery and had rejected it for cause. He lists Naegeli's objections as: (1) lack of constancy of some of the characters Mendel had chosen for study; (2) wide variations in individual ratios as reported by Mendel; (3) lack of evidence in Mendel's paper that reciprocal crosses did in fact give identical results; (4) lack of evidence that other species would give results like those in peas.

Of these, the last was clearly supported by Mendel's own reported failure to confirm this theory with Hieracium crosses, although he did inform Naegeli that crosses in four other genera gave results like those in peas. The other three do not establish Weinstein's main point that Naegeli had appreciated Mendel's theory and rejected it for those reasons. Although Weinstein gives good reasons for believing that Naegeli could understand Mendel's main point as Naegeli seems to have understood a similar view of Naudin which lacked the numerical evidence of segregation, it still remains unexplained why, if Naegeli thought Mendel's points important, he did not say so in print and state his objections there. As it is, our knowledge of Naegeli's views on this matter is based mainly on inference and on a letter to Mendel which is known only through Mendel's reply to it.

Correns, who was a pupil of Naegeli's, had this to say (1905, p. 189) :

> It was doubtless the sharp antithesis between variety [*Elementarart* of de Vries] and race [which de Vries referred to as "variety"] and the view that races in nature were not capable of competition which led Naegeli, whose interest was directed to the problem of species formation, not to value highly as we do now the researches of Mendel which were carried out on typical races.

In addition, Mendel's idea of segregation of alleles was entirely new, and the rather abrupt and succinct way in which Mendel introduced it did not recommend it to a man like Naegeli, who had fixed his mind on what he considered to be larger problems. There is probably some truth to the explanation often offered, that Mendel was dealing with the minor tactics of evolution, and only indirectly at that, at a time when biologists had their thoughts and ambitions focused on the kind of grand strategy represented by the *Origin of Species.*

Bentley Glass (1953, p. 157) believes that "Naegeli completely failed to appreciate the significance of Mendel's discoveries." He points out that Naegeli was an idealist in philosophy and a believer in an "inner perfecting force." This led him to reject external forces such as natural selection as not being of primary importance in evolution and to admonish Mendel by stating that his results were merely "empirical," not "rational" or derived from some over-all principle such as Naegeli's "inner force."

Other explanations for the neglect of Mendel's work have been offered by Alfred Barthelmess (1952) and E. B. Gasking (1959), among others. These include biologists' preoccupation with speculation concerning Darwin's theory of evolution, obscurity of the journal in which Mendel published his results, and similar reasons based on external circumstance. None of these seems very convincing when placed beside the fact that some of the leaders in the field like Naegeli and Kerner knew of the paper but did not review it or discuss it. To them, Mendel was probably an outsider, an amateur. He was referred to as the "abbot of Brünn" and Barthelmess suggests that the anticlerical attitudes of the 1870s and 1880s may have had something to do with the failure of those references to arouse a response among botanists. It is more likely that nineteenth-century biologists were unaccustomed to thinking in the statistical manner which Mendel introduced in the study of hybridization. They were not looking for the kind of theory which Mendel proposed.

Mendel never called himself to the attention of the scientific world. His letters to Naegeli reveal a humble person more interested in testing the truth of his principles than in propagating them publicly; while Naegeli's advice to Mendel sounds like that of a highly placed professional patronizing an amateur. A fact influencing Mendel to withhold further publication and eventually to cease scientific work in this field may well have been his defeat by the unsuitable Hieracium, and the failure of Naegeli to understand the principles derived from peas. There is even a hint in Mendel's letter to his brother-in-law Alois Schindler, dated March 26, 1868 (cf. Iltis, 1932, p. 239), that he welcomed election as abbot in spite of the inroads it would surely make upon his free time for research.

While judgment of Mendel's place in the history of science must rest in the main on the evidence of his hybridization paper of 1866 and of his ten letters to Naegeli (1867–1873), there are some sparse details of his life and work which help us to form a more adequate, if still incomplete, picture of the man.[6] Very few of these details come from Mendel himself, for he kept no diary and wrote little about himself. Although he was a priest and a teacher, the investigation of

[6] See biographical outline at the end of this chapter.

natural phenomena seems to have been the dominant interest of his life, at least until his election as abbot at the age of forty-five. Some of this interest, first expressed in gardening and horticulture, derived from his peasant origin and boyhood work on his father's farm. He is said also to have bred mice and birds, but there is only indirect evidence of this.

His devotion to botanical observation and experiment extended throughout his life. As he grew older and more corpulent, collecting and other field activities were succeeded by more sedentary work in the garden and orchard, increasing attention to his apiary and bee-breeding, and especially to his long-continued and carefully recorded daily meteorological observations. There was always a practical side to his interest in these matters, as in the improvement of vegetable crops and fruit for which he made hundreds of grafts and crosses. From one of his trips to Italy he brought back seeds of Italian grapes which he planted as part of a plan for rehabilitating, at his own expense, a barren hillside near the monastery. He was an active member for many years of the Silesian and Moravian Agriculture Society, serving on the executive committee and as its examiner for fruit and vegetable growing.

There can be no doubt, however, that the prevailing temper of his mind and probably his primary motivation was scientific. This comes out clearly in the hybridization paper, in the letters to Naegeli, and in his chief published contribution to meteorology (1871). The latter was a description of the tornado which struck Brünn on October 13, 1870; but it was a description accompanied by a new interpretation of the cause of tornadoes as vortices engendered by encounters between conflicting air currents. This paper, too, seems to have been overlooked even by those who many years later developed a similar explanation. He was a founding member of the Austrian Meterological Society and his reputation as a meterologist responsible for the recording stations at Brünn extended far beyond his own province and nation.

It is clear from Mendel's annotations in several of the works of Darwin and in other scientific books that he purchased for the monastery library, as well as from published acknowledgments to botanical colleagues and forerunners, that he read widely in scientific literature. While still engaged upon his pea experiments, in 1862 he joined with several like-minded colleagues in Brünn to found the Natural Science Society before which, in 1865, he presented the account of his hybridization research. He took an active part in its affairs, becoming vice chairman in 1869.

An important facet of Mendel's scientific background remains quite obscure. This was his knowledge of experimental physics and of the mathematical ideas and methods which necessarily accompany it.

It was the professor of physics, Franz, at Mendel's preparatory school who recommended him for further study and to the Brünn cloister. Mendel later studied physics at the University of Vienna, one of his professors being Christian Johann Doppler (discoverer of the wave-motion effect that bears his name). Mendel also served for a time as assistant demonstrator at the Physics Institute while a student at Vienna. He was unsuccessful in the physics part of the examination for a teacher's license, yet he taught experimental physics (as a "supply" teacher) at Brünn Modern School from 1854 until 1868 and was apparently successful at that (*cf.* also Dunn, 1965).

It would be interesting to know more about how this background influenced Mendel's practice of thinking in quantitative terms and of giving a clear statement of questions to be answered by experiment which were such prominent features of his method of study. It is unlikely that these qualities owed much to botanical study. In fact, the numerical statistical features of the hybridization paper may have been just what led nineteenth-century botanists not to take it seriously.

Although Mendel's devotion to science was in evidence throughout his life, his active investigations lasted only from 1856, when the pea experiments began, to 1870 or thereabouts, when the failure of the Hieracium experiments and his increasing administrative and public responsibilities put an end to experimental work. Not only did he become the administrative head of the monastery, he also served on many committees, work that included the directorship of an institute for deaf mutes and the chairmanship of a mortgage bank. Moreover, in later life, he traveled abroad frequently, sometimes to visit botanical gardens or to attend congresses and exhibitions, as in Kiel, Paris and London, more often, perhaps, to rest and escape from the burdens of administrative work. His visits to Switzerland and to Italy probably fall in the latter category.

All in all, the picture emerges not of an obscure priest, as some accounts have had it, but of a many-sided man who not only was competent in the restricted field of research in which he was a true creator, but also bore an important part in the life of his town and of his nation.

How it came about that his work had no effect on the progress of science until thirty-five years after its publication is a question to which no single or simple answer can be given. Mendel's personal qualities contributing to his lack of recognition were, paradoxically, those which we can recognize today as having been partly responsible for his achievement. His native modesty and his reticence (twenty years of scientific work compressed into four brief papers) were expressions of the same attributes of mind and character which led him to strip a problem to its bare essentials, and this simplification would lead to its solution. It should not be overlooked that he did not propose to solve the whole problem of heredity, nor of the origin of differences between species, nor of evolution, but to see whether there was a "generally applicable law governing the formation and development of hybrids."

Biographical Outline of Mendel's Life

I have refrained from repeating here details of Mendel's life as recounted by Correns (1922), Iltis (1924 and 1932),

Richter (1943), Dodson (1955) and Krumbiegel (1957). The following dates from these sources will serve as an outline.

1822 July 22	Johann Mendel, born in Heinzendorf, Moravia, the son of Anton Mendel, peasant.
1840 August 7	Graduated from the Gymnasium in Troppau.
1843 October 9	Entered the Augustinian monastery of St. Thomas, the Königinkloster in Altbrünn, as novice, taking name of Gregor.
1848 August 6	Ordained a priest.
1849 September 28	Appointed substitute teacher in Znaim high school (mathematics and Greek).
1850 November	Failed examination for appointment as regular high school teacher.
1851 April 6– July 8	Substitute teacher in Olmütz Technical School.
October 27	Left Brünn to begin studies at University of Vienna — four terms — experimental physics (Prof. Doppler), physical apparatus and mathematical physics (Prof. Ettinghausen); zoology (Prof. Kner); botany (Prof. Fenzl); plant physiology and paleontology (Prof. Unger) — also further mathematics and general paleontology.
1853 January 5	Membership in Zoological-Botanical Society of Vienna.
Summer	End of summer term — returned to Brünn.
1854 April 5	Letter to Vienna Zoological-Botanical Society about damage by the pea-weevil, *Bruchus pisi* (first indication of interest in peas).
May 26	Supply teacher at Brünn Modern School (Staatsrealschule) in physics and natural history — a position held until 1868.
1856 May 5	Took examination for teaching certificate in physics and natural history but did not qualify.[7]

[7] For newly discovered details see J. Krizenecky (1963); Mendel withdrew because of illness.

Spring	Began experimental crossing of pea varieties.
1862	Became a founding member of the Brünn Natural Science Society (Naturforschenden Verein in Brünn). Visited Paris, London and Rome.
1865 February 8	First lecture on pea experiments to Brünn Natural Science Society.
March 8	Second lecture as above.
1866	"Experiments on Plant Hybridization" published in Volume 4 of the *Proceedings* of the Natural Science Society. Began experiments with other plant species.
December 31	First letter to Naegeli.
1868 March 30	Elected abbot of Monastery of St. Thomas.
1869	Lecture on Hieracium experiments at Natural Science Society.
1870	Hieracium paper published in the *Proceedings* of the Society.
November 9	Lecture to Natural Science Society on the Brünn tornado of October 13, 1870 (published in the *Proceedings*, 1871).
1871 September	Visited Dresden, Hamburg, Kiel (Congress of Apiarists).
1873	Last letter to Naegeli.
1874–1883	Struggled as abbot against paying taxes imposed on the Monastery.
1884 January 6	Died of Bright's disease.[8]
January 9	Buried in Brünn Central Cemetery.

[8] For details see J. Sajner (1963); prescriptions during last illness and diagnosis: nephritis with generalized edema.

– Chapter 2 –

MENDEL'S PREDECESSORS:
PLANT HYBRIDIZERS OF THE
EIGHTEENTH AND NINETEENTH CENTURIES

OF THE SEVERAL LINES of development which came together in giving rise to modern genetics, the main or direct
one was represented by the experimental botanists and especially by the plant hybridizers. The essential questions concerning the mechanism of reproduction in flowering plants
and the nature of species in plants could not be settled by
observation or description alone but could, and did, yield their
secrets to experimental study; and it was from experimental
breeding that the principles of genetics were discovered. It is
possible to summarize very briefly here the work of the early
plant hybridizers because of the thorough historical treatments
of their work by H. F. Roberts in his *Plant Hybridization
before Mendel* (1929), and by Conway Zirkle (1935). Much
the same ground is covered with less documentation by Hans
Stubbe in his *Kurze Geschichte der Genetik* (1963).

A sexual process in plants was unknown to Europeans until
the end of the seventeenth century. Then the English physician Nehemiah Grew in his *Anatomy of Plants* (1782)
explained the function of pollen. This seems to have been
recognized in England and America even before the first
experimental proof was given by Rudolf Jakob Camerarius
in Germany. His *De Sex Plantarum Epistola*, published in

1694, contained convincing demonstrations that the pollen supplies the male elements and seed-bearing flowers or plants, the female elements, and that the union of the male and female elements is essential for the formation of the new individual in plants as in animals. Certain exceptions to this rule, as when a female plant isolated from males of its own species nevertheless produced some seeds, led Camerarius to suspect that pollen from another species might be responsible. Thus, to quote M. J. Sirks (1956, p. 68): "With this, Camerarius began the long chain of people who became interested in this problem of hybridization and its consequences."

Artificial fertilization of the blossoms of the date palm by pollen from a male tree was practiced in Babylonian times, but this empirical and partial knowledge of sex in plants was not generalized or used in the breeding of other plants. Its rediscovery and elaboration occupied the attention of a succession of botanists in Europe and America for 200 years. Practical use of the new knowledge was made by Thomas Fairchild in England, who, in 1717, produced a sterile hybrid from two species of Dianthus (Carnation × Sweet William). In America, Cotton Mather had reported (1716) on crosses between squash and gourd (Zirkle, 1935). Several pupils of Linnaeus, stimulated by the work of Camerarius, occupied themselves with sex and plant hybrids in the mid-eighteenth century, and in 1760 Linnaeus responded to the prize offer of the Imperial Academy of Sciences in St. Petersburg by submitting a *Disquisitio de Sexu Plantarum* which won the prize (*cf.* Roberts, 1929, pp. 18–19). Evidence of the sexual process was derived from hybridization and naturally occurring hybrids between different plant species. These observations so shook Linnaeus' faith in the unchangeableness of species that he subsequently struck from a new edition of his *Systema Naturae* the famous sentence, "Species tot numeramus, quot diversae formae in primitione sunt creatae" ("We count as many species as different forms were created in the beginning"; cited from Stubbe, 1963, p. 76).

However, the real founder of modern studies of sex in plants and of scientific plant breeding was Josef Gottlieb Kölreuter. If there is one person to be chosen as the direct scientific ancestor of Mendel, de Vries, and the other founders of genetics, he must be Kölreuter, for it was Kölreuter who, in 1760, in a deliberately planned program produced the first plant hybrid resulting from an experiment.[9] This hybrid from *Nicotiana paniculata* × *Nicotiana rustica*, as well as its reciprocal, was repeatedly produced. Many other hybrids between plant species were analyzed in backcrosses with parent species and in F_2 generations in a modern manner. The functions of pollen and ovule were exhaustively studied, so that controlled and accurate experimental analysis of F_1, F_2, and backcross generations became possible. These indicated that hybrids did not breed true but gave rise to different degrees of mixture of the characters of the parents and of the hybrids, as well as to new combinations.

As Zirkle (1935) has shown, interest in plant hybrids was by no means absent before the publication of Kölreuter's main work in 1761, and in general had a more scientific form than did the theories about animal hybrids which often still reflected primitive superstitions.

Kölreuter's brilliant work on hybridization seems not to have had much influence on his contemporaries or immediate successors. His foundation of knowledge of the sexual process in plants was extended by Christian Konrad Sprengel, whose book of 1793, *Das entdeckte Geheimnis der Natur im Bau und in der Befrüchtung der Blumen*, presented proof that plants with certain kinds of flowers are naturally cross-fertilized in nature. Roberts (1929, p. 82) says of this work:

It signified that the bringing together of combinations of parental characters is the rule rather than the excep-

[9] This statement is subject to some qualification since Ernst Lehmann (1916) has called attention to experiments by M. Guyot who, in 1752, made crosses among varieties of Ranunculus and Auricula differing in flower color.

tion in nature and that therefore the breeding of new types in the plant world may be said to be going on all the time. It remained for Darwin to show how the results from such perpetual crossings are limited and held in check by the operation of natural selection. At all events, Sprengel's discoveries at once disclosed at least an important reason for diversity, for so many variations in nature. . . .

Early in the nineteenth century, serious and extensive work on hybridization as a means to improve domesticated plants was undertaken in England by T. A. Knight and William Herbert.[10] In 1823, there appeared a paper by Knight on crossing experiments with the garden pea, motivated by purposes similar to those of Mendel but lacking the precise framing of questions which distinguished Mendel's work. Knight observed dominance of seed color in F_1 and recovery of both parental colors in F_2, but he did not report ratios. In 1822, another English horticulturist, John Goss, published an account of crosses among pea varieties which came even closer to Mendel's results, since he noted dominance in F_1, segregation in F_2, and found that recessives bred true in F_3 while seeds with the dominant character in F_2 again produced both parental colors. But neither Knight nor Goss, nor Alexander Seton, who also (in 1822) obtained similar results with cotyledon color differences in peas, recognized the need for counting the different types, nor did they propose any systematic interpretation. Much later, between 1868 and 1872, an English seed breeder, Thomas Laxton, without knowledge of Mendel's work, confirmed the results of Knight and Goss and Seton and noted also the assortment of two pairs of contrasted characters, but again without discovering any rule.

In France, also, hybridization was under active study. Augustin Sageret, an agronomist, published an account of crosses between muskmelon and cantaloupe in which for the

[10] The following four paragraphs are based on Roberts (1929), which contains detailed references to the original literature.

first time the separate characters of the two parents were listed in contrasted pairs. In reporting evidence on descendants of hybrids, Sageret (1826, p. 302) made the important point that:

> in general, the resemblance of the hybrid to its two ascendants consisted, not in an intimate fusion of the diverse characters peculiar to each one of them in particular, but rather in a distribution, equal or unequal, of the same characters.

The most extensive publication on hybridization during the period after Kölreuter and before Mendel was a work by Karl Friedrich von Gaertner. First submitted as an essay in competition for a prize offered by the Dutch Academy of Sciences in Haarlem, it was published in Dutch (1838) and then amplified and revised in the definitive German edition of 1849. It reported the results of thousands of experimental hybridizations. Gaertner emphasized the equivalence of reciprocal crosses, the fertility of many hybrids which in further breeding produced offspring resembling the original parents, and in general the increased variability after crossing. Among his predecessors, Mendel singled out especially Gaertner's "valuable observations."

The person most often referred to as the immediate forerunner of Mendel was Charles Naudin of the Museum of Natural History in Paris. In 1863, in response to a prize offer of the Paris Academy of Sciences, he submitted a paper on hybridization which took the prize (*cf.* Roberts, 1929, p. 131). In it, he stated a

> . . . law of disjunction of two specific essences in the pollen and ovules of the hybrid. A hybrid plant is an individual in which are found two different essences, having their respective modes of development and final direction, which mutually counter one another, and which are incessantly in a struggle to disengage themselves from one another.

T. H. Morgan has estimated the position of Naudin in the ancestry of genetics as follows (1932, p. 264) :

Naudin stated explicitly that in the second and later generations there is a mixture of forms, including some which are like the original parents and others that approach these in various degrees. Then follows his most important deduction, namely, that the second generation results find their explanation in the disjunction of the two specific essences derived from the parents in the ovules and in the pollen of the hybrid. Here we have a highly significant contribution, for, not only did Naudin see clearly that the results are explicable on the principle of disjunction (or, as we say now, segregation), but that this taking place both in the egg and in the pollen, gives the kinds of characters that appear. So important historically is this fact that there should be included his specific statements showing that he had a perfectly clear idea as to how disjunction accounts for the diversity in the second generation. If, he says, a pollen grain bearing the characters of the male parent meets an egg of the same kind, a plant that is a reversion to the paternal species will result; similarly for the maternal species. But if a pollen grain of one kind meets an egg of the other kind, a true cross-fertilization takes place like that of the first generation and an intermediate form will result. It will be agreed, I think, on all hands that this was a brilliant interpretation of results based on first-hand experience. It falls short of Mendel's work in two or three important aspects: (1) the failure to put the hypothesis to a test by back-crossing; (2) the failure to see what the numerical results should be on the basis of disjunction of the elements in the hybrid. His use of the words "disordered variation" in the F_2 and later generations brings out the essential difference between Naudin and Mendel. It is the orderly result of disjunction or segregation that is the important feature of Mendel's work; and finally, the clearness with which Mendel stated and proved the interrelation between character-pairs in inheritance, when more than one pair is involved, places his work distinctly above everything that had gone before.

Although several predecessors noted external features similar to those from which Mendel deduced his rules, none of them in fact anticipated his biological interpretation. The same is true of several others who performed their experiments after Mendel's paper had been published, but in ignorance of it. One of these was Charles Darwin (1868, Vol. II, p. 46) who crossed the common snapdragon with the form having peloric flowers and in "two great beds of seedlings not one was peloric." The normal-flowered hybrids when crossed among themselves produced thirty-seven plants with peloric flowers out of 127—a sufficiently close approach to the one-quarter expected on Mendel's theory. But Darwin, says Zirkle (1951a, p. 50), who called attention to this experiment, "made nothing of it." The same was true of other post-Mendelian observations by the Swedish plant breeders P. Bolin and H. Tedin in 1897, and by the German breeder Wilhelm Rimpau in 1891 (*cf.* Stubbe, 1963, pp. 201, 202). Roberts (1929, pp. 276–283) analyzed the results of crosses between winter and summer wheat varieties begun in 1899 by W. J. Spillman of the U.S. Department of Agriculture. Typical Mendelian results for at least three characters were reported in statistical form but were not analyzed for ratios which Roberts showed were close to 3:1 in most cases, while another character difference gave a good 1:2:1 ratio in the F_2 generation.

Spillman, although he did not reach Mendel's explanation, recognized that the array of characters found in the F_2 generation following a cross "include all possible combinations of the characters of the two parents" (Spillman, 1901, p. 89). He also noted, as a rule, that the most frequent type of F_2 was the same as the uniform F_1 type. This important paper of Spillman's shows that in his recognition of the requirements for quantitative observation of single traits and their combinations and in the results obtained he came very close to a Mendelian interpretation. However, he made no such claim and in fact did not know of Mendel's paper when he gave his report in November, 1901.

Thus, for nearly a century, observations were available which might have yielded Mendel's principles. They did so only in the mind of Mendel, and after an interval de Vries, Correns, and Tschermak rediscovered the principles. The question why Mendel's predecessors who had also observed the offspring of hybrids did not discover these rules cannot, of course, receive any definite answer. An essential difference between Mendel and them is that he was looking for such rules and they were not. Thinking in terms of inherited units which obeyed statistical laws was obviously not common in the nineteenth century. Living units themselves, as the next chapter will show, were familiar to biologists. Behavior according to a binary system which generated regular ratios was the new invention.

– Chapter 3 –

IDEAS ABOUT LIVING UNITS:
FROM SPENCER TO JOHANNSEN
(1864–1909)

By the period 1935–40, the three main directions in which genetics was to develop had become clearly evident and the basic questions of each had been outlined. The transmission mechanism of genes in chromosomes had been sufficiently elucidated to serve for many years to come as the basis from which attacks on the other two problems—evolution and individual development—could be launched. In fact, this had been the case almost from the period when the general truth of the principle of segregation had been recognized. That period, say from 1900–1906, can now be seen as marking the chief break in the continuity of ideas about the transmission system of heredity. What had happened prior to that time— except, of course, for the original publication of Mendel's pea experiments—had little connection with the later course of development of these ideas. There was no such sharp break in studies of evolution or of development; in fact, many students of the latter field would say that it has not yet been affected in a major way by what happened in 1900. Views as to the mechanism of evolution began to change gradually in the late 1920s and early 1930s.

It is, of course, not surprising that the effects of the break should be felt first in the territory, so to speak, in which it occurred, that of the transmission mechanism of heredity. It is an interesting reflection, from the vantage point of 1965, that that "territory" had not been recognized as such until after the break had occurred. Heredity, with the definiteness given to it by modern genetics, now seems to be a well-marked field, but in the nineteenth century this was not so. Then, efforts to develop general theories of heredity were judged primarily by their applicability to problems of variation, evolution, and development, and we see signs today, after nearly 100 years, of a return of that attitude. Then it was due to the absence of precise questions about the transmission mechanism or to a lack of recognition of the territory. Now genetics threatens to disappear as a separate field because the solution of some of its basic problems has led to its absorption into the general fabric of biological knowledge. The gene, whatever its fate as an "ultimate" unit in the structure and functioning of living material, has become a basic part of the description of that organization.

The nature of the radical change that occurred around 1900 can be studied by examining some of the questions that were of chief interest to biologists in the immediately preceding period. Three main sets of questions can be distinguished: those concerning the nature of species differences and the effects of hybridization on domesticated plants; those having to do with morphology, development, and cytology, chiefly in animals; and those specifically concerned with theories of heredity. The latter took the form of ideas about living units as elements of hereditary transmission and as guides of the development of form and function. Needless to say, all these questions were directed toward general questions raised as the result of the intense interest in the mechanism of evolution evoked by the publication of *Origin of Species* in 1859.

I shall deal first with the last set of questions, and shall consider the history of ideas concerning living units in the nineteenth century and the state of such theories today. The source of ideas about living units could, of course, be traced to much earlier times. Jean Rostand has made an attempt of this sort in his "Esquisse d'une histoire de l'atomisme en Biologie" (1956)[11]; but the particles imagined by Buffon, Diderot, and others in the period before experimental study began were purely speculative and did not lead to the discovery of the true *atomisme* which Mendel proved. The "elementary particles" of Maupertuis (1751), however, were invoked to account for the segregation of a dominant gene for polydactyly in a human family which he followed through several generations. Maupertuis, as Glass (1947) has shown, was on the right track, although he seems to have had no immediate followers.

Beginning with the "physiological units" on which Herbert Spencer in 1864 based his view of the organization of protoplasm, many thinkers arrived by similar means at similar conceptions. Darwin's "gemmules," Naegeli's "idioplasm," the hierarchy of biophores and ids which Weismann conceived as arranged in "idants" corresponding to visible chromosomes, all belong in this succession, following which, in 1909, Johannsen coined the term "gene." The question is whether the nineteenth-century units represented the same idea as that contained in our present term.

The living units of Spencer (1864) and of Darwin (1868) were invented to explain facts of reproduction, and, especially, of differentiation and development. We should remember that seventy-five years later theories of individual development had still not been satisfactorily accommodated by the theory of the gene. It was especially the repetition of all the details of structure in parts such as a limb or a tail which had been lost and then regenerated that called forth Spencer's idea of physiological units in the remainder of the body which "possess the property of arranging themselves into the special structures of the organisms to which they belong." It is suggestive that the idea of physiological units appears

----

[11] References to literature cited in this chapter will be found in Barthelmess (1952).

first in the *Principles of Biology* in the section on "Waste and Repair." According to Spencer:

> The germ cells are essentially nothing more than vehicles in which are contained small groups of the physiological units in a fit state for obeying their proclivity towards the structural arrangement of the species they belong to.

He imagined them to be units above the level of inorganic molecules and below the lowest level of morphological organization, the cell. The units within any one species were thought to be all alike and each to possess the "polarity" for arrangement into the structures of that species. They could thus be vehicles of transmission as well as directors of development; and since they were assumed to be self-reproducing and to circulate throughout the organism, modifications of bodily parts could result in modifications of the units and thus account for inheritance of acquired characters. This in the mid-nineteenth century was an essential requirement of a theory of heredity.

Darwin's "Provisional Hypothesis of Pangenesis" was set forth in Chapter 27 of *Variation of Animals and Plants under Domestication*. Although published in 1868, four years after Spencer's theory, it does not base itself on the earlier theory, but was Darwin's response to the great variety of facts on variation which he had collected and studied (Vol. II, p. 349):

> In the previous chapters, large classes of facts, such as those bearing on bud-variation, have been discussed; and it is obvious that these subjects, as well as the several modes of reproduction stand in some sort of relation to one another. I have been led, or rather forced, to form a view which to a certain extent connects these facts by a tangible method.

These were facts of asexual and sexual reproduction which he considered to involve essentially the same principles of regeneration, development, graft hybrids, inheritance, effects of use and disuse, reversion and the transmission of characters in latent form. The hypothesis which was to connect these facts assumed that the units known to increase by self-division, *i.e.*, the cells,

> . . . throw out minute granules which are dispersed throughout the whole system; that these supplied with proper nutriment, multiply by self-division and are ultimately developed into units like those from which they were originally derived. These granules may be called gemmules.

These were thought to be extremely small and numerous, circulating in the body and collected into sex cells in which they could be transmitted in a dormant state.

Francis Galton by a simple transfusion experiment with rabbits disproved that part of Darwin's hypothesis according to which gemmules were supposed to circulate in the body and then be collected in the germ cells. However, this did not seem to Darwin to be a fatal objection (1876, Vol. II, p. 350) :

> I certainly should have expected that gemmules would have been present in the blood, but this is no necessary part of the hypothesis, which manifestly applies to plants and the lowest animals.

In its most general form it would in fact have been impossible to disprove Darwin's hypothesis, the essence of which was that new individuals are generated by invisible units rather than by the reproductive organs or buds (1876, Vol. II, p. 398) : "Thus an organism does not generate its kind as a whole but each separate unit generates its kind." These units remained purely hypothetical and this accounts for the withering away of pangenesis as Darwin proposed it. The word itself, however, lived on in the new and different meaning given to it some twenty years later by de Vries in his *Intracelluläre Pangenesis.*

Francis Galton (1822–1911) enters the history of genetics at several points, but first as having given experimental evidence against the transport feature of Darwin's hypothesis of pangenesis. Transfusions of blood were made between rabbits of different colors. Subsequent offspring of these revealed no evidence of having received foreign gemmules. But Galton retained the gemmule part of the hypothesis by assuming that the whole collection of units representing cells of different kinds constituted a hereditary material known as the "stirp" which was distinct from the body cells. The stirp of the child was thought to derive directly from the stirp of each parent, but the bodily characters of the child express only a portion of his own stirp, *i.e.*, those units which become active, while other units remain latent in the stirp but are still transmitted to the next generation.

This sounds very much like Weismann's concept of distinction between an immortal germplasm and a mortal somatoplasm. Galton may well have been the first to express such a distinction. However, Weismann did not concede Galton's priority, for eventually, in order to explain some cases of the inheritance of acquired characters, Galton had to assume that some cells give off gemmules which get into other cells and even into germ cells. Moreover, said Weismann, the true germ plasm theory is independent of any assumption about sexual reproduction, while Galton's stirp is limited to bisexual forms.

It was clear to Johannsen, however, that the essential idea had come from Galton (1875), as shown by the following passage, which I have translated from Johannsen's stimulating and prophetic book, *Om Arvelighed og Variabilitet* (1896):

> In the two preceding chapters we have had to dissociate ourselves from Weismann's speculations; now we must for the third time warn against "Weismannism," which here shows its dogmatic nature in its most beguiling [*besnaerende*] form, namely as the much discussed theory of the Continuity of the Germplasm. The sound conceptions on which this remarkable theory is based are quite old, and so far as we know were first put in a clear form by Galton some 20 years ago, in connection with some now abandoned ideas of Darwin.

Johannsen then described Galton's stirp theory as the direct ancestor of the germ-plasm theory, attributing to Galton the essential idea that the sex cells—egg and sperm—are independent of the body cells, and have their chief function in inheritance as producers of the germ cells of subsequent generations. As a collective name for the elements responsible for transmission, Galton coined the term "stirp" (from the Latin for "root," *stirpes*). According to Galton some of the elements ("germs") produce the body, others produce more germs to form the stirp of the offspring. Hereditary continuity is thus from stirp to stirp, and not from body to body, and this was a new idea, as Johannsen saw. Johannsen's views, which were expressed later in his fundamental conceptions of pure lines, phenotype and genotype and gene, owed more to Galton than to Weismann. Even when we discount Johannsen's aversion to Weismannism as "abstraction and speculation," we must admit that his points were well taken.

There is no doubt that Galton stimulated Weismann to think harder and to refine his speculations about material units of transmission. The versatile Galton had the same effect in many fields. In genetics none of his ideas were incorporated as such into durable theories of heredity, but the statistical methods developed from his first attempts to deal with problems of blending inheritance proved to be extremely valuable in leading to the rejection of many hypotheses, including his own theory of ancestral inheritance.

In the meantime (1884) Naegeli developed his *Mechanischphysiologische Theorie der Abstammungslehre.* Its central feature was that heredity was transmitted by a substance—"idioplasm"—carried by the germ cells but diffused as a network throughout all of the cells. Its elements were complexes of molecules—"micellae"—grouped into units of higher order which during development determined the differentiation of cells, tissues, and organs. These units, too, were based on theoretical speculation, but in the same year, 1884, Oskar Hertwig seized upon Naegeli's idea and for the first time attempted to identify the hereditary substance with the chromatin of the nucleus. A. von Kölliker, Eduard Strasburger, and Weismann quickly arrived at the same view, and what could be called the "nuclear theory" of heredity came into being.

It was August Weismann (1834–1914) whose theoretical insight and ingenuity brought this view to prominence even before the behavior of the nuclear material was elucidated. In Weismann's theory, "idioplasm" became "germ plasm" and this concept of the physical basis of hereditary continuity exerted a strong influence upon biological thinking which has persisted to this day. Although he had expressed views about the physical basis of heredity as consisting of nuclear elements as early as 1883, Weismann's mature statement of his theory was given in his famous book *Das Keimplasma*, which appeared in German in 1892 and in English (*The Germ Plasm, a Theory of Heredity*) almost simultaneously, in 1893.

By that time Weismann had available to him another brilliant and prophetic formulation of a theory of heredity, that which the Dutch physiologist de Vries had published in 1889 under the title *Intracelluläre Pangenesis* (*cf.* de Vries, 1910).

De Vries, a student of Julius Sachs at Würzburg, became professor of botany at Amsterdam in 1878 at the age of thirty. In the early 1880s, he conceived a mechanism for inheritance which bore a closer resemblance to the system as we view it today than did any other pre-Mendelian system. De Vries postulated living, self-replicating units, "pangenes," which provided the model for the gene. Later Johannsen took the term "gene" as well as the concept from de Vries, who paradoxically had borrowed the name from Darwin's pangenesis but had given it an entirely different meaning which he developed fully in his book of 1889. It was this imaginative model which led de Vries to the experimental breeding work on which his later mutation theory was based, and to the analysis of the inheritance of discontinuous characters in fifteen different plant genera. This work was carried out between 1892 and 1899. There is no doubt that in this period de Vries independently discovered Mendel's law of segregation and verified it in many plant species.

The genesis of de Vries' pangene idea is to be found in his

interest in evolution and in what he called "the species problem." In the introduction to *Intracellular Pangenesis* he says:

> If one considers the species characters in the light of the
> doctrine of descent, it then quickly appears that they are
> composed of separate more or less independent factors.
> Almost every one of these is found in numerous species,
> and their changing combinations and association with
> rarer factors determine the extraordinary variety of the
> world of organisms . . . These factors are the units which
> the science of heredity has to investigate. Just as physics
> and chemistry are based on molecules and atoms, even so
> the biological sciences must penetrate to these units in
> order to explain by their combinations the phenomena of
> the living world.

Many of the features of de Vries' hypothesis sound remarkably modern. The existence of a multitude of kinds of invisible
living units in all cells, not all alike as in previous conceptions
but representing separate hereditary predispositions (*anlagen*)
and concentrated in the nucleus, from which representative
particles entered the cytoplasm to influence reactions there;
the properties of self-replication, accompanied by occasional
errors like mutation which gave rise to the variety of pangenes—this sounds very much like a modern description of
genes. Especially prophetic was de Vries' view of a species
as a specific assemblage of individual characters subject to
recombination.

The hypothesis was a great feat of imagination and erudition, but because of its fundamentally deductive character it
suffered the same disability as its predecessors: it did not
include either a prescription or a clear possiblity of proof or
disproof by experiment, although de Vries realized that further
study of it called for crossing experiments. The difference between de Vries' pangenes of 1889 and the concept of the gene
developed by Johannsen twenty years later was the absence
from the former of the concept of alternative states, or alleles.
This, as we know, was derived by Mendel directly from the
behavior of the two forms of a gene in segregation; the only

genes whose existence could be inferred were those of which an alternative form had arisen by mutation.

But if the existence of segregating elements could not have been predicted from de Vries' theoretical model, it served to guide de Vries by way of the study of mutation toward experimental plant breeding and thus to the independent discovery of the principle of segregation. In this sense de Vries' hypothesis of intracellular pangenesis might be called the starting point of modern genetics (*cf.* Stomps, 1954; Heimans, 1962), and he more than any other should share with Mendel the credit for the foundations of the science. He himself did not lay claim to this position and appeared to set more store by his mutation theory than by his discovery of basic principles of heredity.

The three chief contributions of de Vries—the ideas of pangenes, of mutation, and of segregation—were so interconnected in his mind as to form parts of one concept. A few citations from de Vries' writings, originally published in 1889–1900, will illustrate this.

First, in *Intracellular Pangenesis* (*cf.* de Vries, 1910, pp. 70–71) we read:

> In the first division we arrived at the conclusion that hereditary qualities are independent units from the numerous and various groupings of which specific characters originate. Each of these units can vary independently from the others; each one can of itself become the object of experimental treatment in our culture experiments . . . The pangens are not chemical molecules but morphological structures, each built up of numerous molecules. They are like life units, the characters of which can be explained in an historical way only.

> . . . We must simply look for the chief life attributes in them without being able to explain them. We must therefore assume that they assimilate and take nourishment and thereby grow, and then multiply by division, two new pangens, like the original one, usually originating at each cleavage. Deviations from this rule form a starting point for the origin of varieties and species.

At each cell-division every kind of pangen present is, as a rule, transmitted to the two daughter cells.

At the end of this summary section occurs the following sentence (p. 74): "In a word: An altered numerical relation of the pangens already present, and the formation of new kinds of pangens must form the two main factors of variability." In a footnote to this sentence we find: "In a note to the translator, the author says: 'That sentence has since become the basis of the experiments described in my "Mutationstheorie." ' "

De Vries' principle of segregation (which he called "spaltungsgesetz" or "loi de disjunction" in his first papers of 1900) and his pangenesis theory were so closely connected in his mind that the first sentences in his paper of 1900, "Sur La Loi de Disjonction des Hybrides" (taken from Aloha Hannah's 1950 translation), were as follows:

According to the principles which I have expressed elsewhere (*Intracelluläre Pangenesis*, 1889), the specific characters of organisms are composed of separate units. One is able to study, experimentally, these units either by the phenomena of variability and mutability or by the production of hybrids.

After citing the results in the F_2 generation of single character segregation in eleven species of plants, he concludes:

The totality of these experiments establishes the law of the segregation of hybrids and confirms the principles that I have expressed concerning the specific characters considered as being distinct units.

It is clear that de Vries was not a "re-discoverer" but a creator of broad general principles and that his work extended that of Mendel in at least three ways: (1) it proved the operation of segregation in a wide variety of plant species; (2) it made material units an essential feature of the theory; and (3) it introduced the concept of mutability as the source of variety in the units. If the stature of de Vries has steadily increased as genetics has developed, it is perhaps in part because genetics grew to resemble the view he entertained in the 1880's—that the variety of the organic world was maintained through the properties and activities of material units which were the prime operators in the transmission of heredity, in evolution, and in development.

I return now to the other great theorist of the pre-1900 period, August Weismann. His influence on the development of genetics lies mainly along another track, one which led to the chromosome theory as it matured in the 1920s and 1930s. To him the importance of the living units he envisaged was that they constituted a hierarchy of ascending size culminating in material bodies which could be seen with the microscope. The concept of the continuity of the germ plasm guided and inspired his thinking as the idea of pangenes had done for de Vries, but it led him toward cytological observation and not toward experiment, as in de Vries' case. He was, however, frustrated by an eye disease, and his energies turned to speculation and constructing models. In his essay of 1883, he began with M. Nussbaum's view of the continuity of the germ cells and developed it into the germ-plasm theory which embodied a new and fruitful view of the process of inheritance. According to this view, the offspring does not inherit its characters from the body of the parent but from the germ cell which has derived its properties and elements from the ancestral germ cells which preceded it. The body, or soma, is viewed as the temporary and mortal custodian of the continuous line of germ cells. Samuel Butler described this idea well: "It has, I believe, been often remarked that a hen is only an egg's way of making another egg." To place chief emphasis on the germ cells and the hereditary material itself was a fundamental advance in the direction leading to modern genetics. Today we say that we do not inherit characters but genes, which in the course of development and in interaction with the environment produce characters. Johannsen's distinction between

phenotype and genotype and Woltereck's conception of the
norm of reaction, now seen as views essential to an understanding of heredity, both trace ultimately to Weismann and
Nussbaum's distinction, and behind them, according to
Johannsen's view, is Galton.

It was the germ-plasm theory which led Weismann to reject
the Lamarckian theory of the inheritance of acquired characters. This had the effect of directing evolutionary thinking
into paths which led toward genetics. It led too to a theory
of an elaborate hierarchy of material bodies, beginning with
"biophores" of molecular order of size, equivalent more or
less to de Vries' pangenes. These in turn were grouped into
"determinants," which as "pieces of inheritance" (*Verer-
bungsstücke*) determined the forms of cells and cell groups
and thus the inherited characters. Determinants were seen as
grouped into a third level of organization in "ids," the name
intended to sound like Naegeli's idioplasm. Each id was
viewed as containing the whole architecture of the germ
plasm. They were arranged in linear sequence to constitute
the "idant," which was Weismann's term for the chromosome.
This view of the newly discovered and named chromosome expressed its function in transmission and led directly to Weismann's assumption of a reducing division. If idants replicated,
as mitosis showed they did, then something must intervene to
hold in check "the excessive accumulation of different kinds of
hereditary tendencies or germ plasms." How Weismann's prediction of a reduction division during meiosis was fulfilled
belongs to the history of cytology. Similarly, the defeat of his
application of the determinant theory to development—the
postulate that mitoses were qualitative, in the sense that determinants were progressively sorted out to different cells and
organs—belongs to a history of embryology.

It was Weismann's insight into basic or elementary processes
and his skill in stating his ideas clearly and forcefully that
gave him the extraordinary influence that he exercised in
the last twenty years of the nineteenth century. Here is an

example which I have translated from his essay of 1883, "Über die Vererbung":

> What one understands in general by heredity is well enough known; it is the property of all organisms to transmit their own character to their descendants.

> But on what does this common property of organisms rest?

> It was Haeckel who first called reproduction a growth beyond the mass of the individual and tried by this to make heredity more comprehensible since he conceived of it as simple continuation of growth. One could easily take this for a mere play on words, but it contains more: yes, when rightly used, this conception even points to the only way, it seems to me, that can lead to understanding.

> One-celled organisms, such as infusoria, multiplying by division, grow up to a certain size and then divide into two halves which resemble each other entirely not only in size but in activity.

> In such one-celled forms we understand thus up to a certain point why the bud is like the parent—it is just a piece of it. Really, the question why the piece must be like the whole leads to a new problem, that of assimilation, which itself awaits a solution. Nevertheless, the undoubted fact is, that organisms possess the ability to take up and to convert into their own substance certain external substances: food.

> Heredity in these one-celled forms derives from the continuity of the individual whose bodily substance increases more and more by assimilation.

> But how is it with the many-celled organisms which do not reproduce by simple fission, and in which the quality of the whole body of the parent is not carried over in the bud?

> In all multicellular animals sexual reproduction forms the basis of their multiplication; none lack it entirely and in the majority it is the only method of multiplication. With them reproduction is connected with certain cells, which one may designate as germ cells, which one may contrast with the cells which form the body, and indeed

must contrast them, for they play a totally different role. They are without importance for the life of their bearers (that is for the maintenance of life), but they alone maintain the species, for each one of them may under certain conditions again develop into a complete organism of the same species as the parental one, having all possible individual characters of the parent more or less expressed. Now how does the transmission of parental characters to progeny occur, how is it possible for one germ cell to reproduce the whole body with all its peculiarities?

If, in this it were only a question of the continuity of substance from one generation to another, this would easily be accomplished, for it could be easily demonstrated in individual cases and made probable in all. But since, as their development shows, there is a more fundamental contrast between the substance or the plasma of the immortal germ cell line and the mortal cells of the body, we can hardly view this fact otherwise than that both kinds of plasma are kept in the germ cell as possibilities, which separate from each other sooner or later in the form of detached cells after the beginning of embryonic development.

Thus we have as well in the reproduction of metazoa as in one-celled forms what is at bottom the same process: a continued division of germ cells, and the difference lies only in this, that here the germ cells do not make the whole individual, but that these are enclosed by many thousands, even millions and billions of body cells whose aggregate first forms the higher unity of the individual. Thus the problem posed above—how does it happen that one germ cell contains the *anlage* for the whole formed complex individual?—must be stated more precisely: How is it that in the higher animals the plasma of the germ cells holds potentia for the somatoplasm, or better, such plasma as is capable of developing into body cells, and especially of quite specific quality?

Those words were written eighty years ago. The dilemma so clearly stated by Weismann is still with us.

The passage above illustrates another attitude which was current at the end of the nineteenth century. "Continuity of substance," transmission, was no great puzzle; differentiation, morphogenesis—that was the real challenge. The same is true today, but with this difference. The nineteenth-century biologists, except Mendel and the "rediscoverers," did not begin to solve the problem of transmission because they failed to recognize its real nature or even its importance. It was only when some biologists were willing to put aside the intractable problem of development and concentrate on transmission that the problem was analyzed and solved.

It is interesting that many of the speculations involving living units, *e. g.* those from Spencer in 1864 to Weismann in 1883, met defeat by failing to account for the observed facts of development and regeneration. Yet the theory of the gene, which also failed to account for development, was accepted because of the compelling nature of the evidence from experimental breeding, backed up by cytological observation. The theory of the gene was concerned only with units as vehicles of transmission and in Morgan's words (1926, p. 26) "states nothing with respect to the way in which the genes are connected with the end product or character." Nevertheless the arguments for the restricted theory were so persuasive that they were accepted by most biologists, in spite of the paradox that the mechanism proposed assumed the same variety of units in all cells although the cells themselves become different.

The break, at 1900, between the earlier period of construction, by deductive reasoning, of speculative schemes to explain heredity and the later period in which theory was based on experimental investigation was also marked by sharp contrast between concepts of the nature of the living units employed in the two periods. Most of the "deduced" units were conceived of as material elements—physiological units, gemmules, pangenes; while those arrived at inductively in the Mendelian scheme did not have this quality. The observed realities in Mendel's experiments were things grossly seen, the differentiating characters themselves. The Mendelian "elements" within the reproductive cells were symbols only, inferred from statistical rules.

There was one man, however, who bridged the gap between

the two periods since he worked by both methods. Before 1899, de Vries had constructed a system of intracellular units which he considered to derive from Darwin's intercellular pangenesis although it was fundamentally different. He also reached, inductively, by planned experiments, the same rules which had been derived by Mendel. None of his contemporaries (he was born in 1848) or fellow-thinkers was able to make this transition. Galton, born the same year as Mendel (1822), Spencer (b. 1820), Weismann (b. 1834), and Haeckel (b. 1834), were perhaps too old when the big break came.

William Bateson (b. 1861), who must in a formal sense be considered the founder of genetics, since he first formulated its principles and gave it its name, had little patience with the kind of speculation which dominated formal or academic biology (but not animal breeding!) in the nineteenth century. His scornful review (1905) of Weismann's *Evolution Theory* (1904) shows what he thought of Weismann's idea of selection among germinal elements:

> Natural selection, being here in her economical mood, is to eliminate the parts which if allowed to develop, would not have pulled their weight. We thus meet a proleptic Natural Selection, dealing in futures—as transcendental a conception as any Nägelian could desire . . . Evolution has passed out of the speculative stage.

And so, Bateson might have added, had heredity.

– Chapter 4 –

PREPARING CYTOLOGY FOR MENDELISM

IT IS CLEAR that for Weismann the same ideas of order should serve for inheritance, evolution, and development. For him as for many nineteenth-century biologists the analysis of the transmission mechanism was not the prime problem; it was rather the behavior of the living units in controlling the differentiation of parts. Of course, these pre-1900 attempts tackled what proved to be the most difficult problem first— that of development. In the nineteenth century, heredity included development, as in fact any complete theory must. Perhaps one should say, more precisely, that the problems of inheritance and development which could be examined separately after the clues to the transmission mechanism were discovered and exploited had at that time not been subjected to this cleaving influence.

We recognize now that a living unit as a means of continuity between generations and as an element of change in evolution has, as its most important property, self-replication with opportunity for mutation; but when it is thought of as being responsible for initiating the development of a part of the organism in the pattern of a predecessor, then the property receiving greatest emphasis will be not specificity of replication but its ability to act as a surrogate in initiating reactions like those which occurred in the ancestor. It was not only nineteenth-century theories which failed in this latter aspect but twentieth-century ones also.

If one focuses too closely or narrowly on theories of heredity, the period between 1865 and 1900 may seem like a gap. Mendel's theory was worked out at the earlier date and Darwin's hypothesis of pangenesis, although first published in 1868, had been in his mind for nearly thirty years. In 1900, one of these theories (Mendel's) was reborn, and by the same token the other was excluded. But the last thirty-five years of the nineteenth century were anything but empty ones to those biologists who, chiefly in a rush of progress in the 1870s and 1880s, brought modern cytology and embryology into being. Between the discovery of cell division and fertilization in the mid-seventies and the elucidation of the reduction division near the turn of the century, there was the same kind of excitement that accompanied the rapid development of the theory of the gene between 1910 and 1925. The earlier period has been described in Arthur Hughes' *A History of Cytology* (1959) and need not be reviewed here in detail. William Coleman (1965) has given an excellent account of the development of ideas, in the period 1840–1885, concerning the nucleus as the vehicle of inheritance.

Thus it was that when Mendel's principles were rediscovered, knowledge about the physical mechanism in the chromosomes was available and the connection was quickly seen by Walter S. Sutton, de Vries, and others. It is clear that the speed with which genetics developed after 1900 was due not only to the ease with which breeding experiments could be used to test theories of heredity but to the adequacy and soundness of the cytological work of the 1870s and 1880s. The substructures proved to be strong enough and broad enough to bear the weight of the great theoretical construction which was to rest upon it. This could not have been the case in 1865, for at that time not even the nature of cells and of cell division was correctly known. Indeed, although the so-called cell theory of M. J. Schleiden and Theodor Schwann dates to 1838, and Rudolph Virchow's famous aphorism "omnis cellula e cellula" to 1855, the cells they talked about were not those which later embryologists, cytologists and geneticists used as operating terms in their theories. An essential conception is the manner of formation of new cells. If they were subject to "free formation" in fluid or by expansion of the nucleus into a new cell which replaced the old one from within—as Virchow thought—they could hardly become the vehicles of the continuity required to produce the patterns of the parent in the offspring or of cellular heredity.

Between 1844 when Naegeli observed nuclear division (but thought it exceptional) and 1880 when general agreement had been reached that cells were formed by division with the nucleus dividing first, a solid observational basis was laid for the interpretation of karyokinesis or indirect cell division. The exact longitudinal division of the chromatin threads with each part replicated was seen as the essential function of the process of mitosis. The chromosomes (so named by W. Waldeyer in 1888) were seen to consist of smaller elements: "chromomeres" and "chromioles." Gustav Eisen, in 1899, counted in a salamander, Batrachoseps, twelve chromosomes, each with six chromomeres, each chromomere consisting of six smaller granules, the chromioles. Thus about 400 elements, he thought, could be distinguished in the haploid male pronucleus. (In 1928, John Belling counted the "ultimate chromomeres" in a lily and found 2000–2500 in the haploid set. The purpose was the same but the techniques of preparation and microscopy better at the later date.) Such counts were certainly not exact but they expressed a view which became more general that chromosomes were individual and highly differentiated structures. The final proof of this was provided later by Sutton in 1903 and by Theodor Boveri in 1904.

It was not merely neglect of Mendel's work which was responsible for the stasis of genetical thinking between 1865 and the 1880s. Biologists were thinking about other matters which proved to be just as important. They were laying the groundwork for a sound biology of reproduction in both plants and animals. Strasburger, Van Beneden, Flemming, Overton, Fol, Boveri and Nawaschin prepared the way by detailed observations of nuclear structures and nuclear division. For an appraisal of that period and of the state of biological thinking on the eve of the Mendelian revelation, one cannot do better than turn to the second (1900) edition of E. B. Wilson's *The Cell in Development and Heredity*. In an appreciation of Wilson by his student, H. J. Muller (1943), Wilson's book and its effect are discussed under the heading "Preparing Cytology for Mendelism." That is what was happening in the 1880s and the 1890s. Wilson's book as first published in 1896 stated the essential questions of heredity and described the state of cytological knowledge. The question was:

> How do the adult characters lie latent in the germ cell; and how do they become patent as development proceeds? This is the final question that looms in the background of every investigation of the cell.

The honest answer in 1900 was *ignoramus*. Inheritance and development still remained as a riddle in their fundamental aspects. But there had been progress. Naegeli in 1884 had published the first systematic attempt to discuss heredity as inherent in a definite physical basis. Inheritance, he supposed, is affected by transmission, not of a cell, considered as a whole, but of a particular substance, the idioplasm, contained within a cell and forming the physical basis of heredity. Naegeli had made no attempt to locate the idioplasm precisely or to identify it with any morphological constituent. But the advance of cytological knowledge led others to locate it in the nucleus, as chromatin.

Oskar Hertwig, Strasburger, von Kölliker and Weismann independently and almost simultaneously (1884–85) came to the conclusion that "the nucleus contained the physical basis of inheritance; and chromatin, its essential constituent, is the idioplasm" (Wilson, 1900). The identification of chromatin with a chemical substance (nuclein, first described by Miescher in 1872) was made by Zacharias in 1881. These events, followed by those of 1884–85 led E. B. Wilson to write in a prophetic passage (1895, p. 4) :

> Now, chromatin is known to be closely similar to, if not identical with, a substance known as nuclein . . . which analysis shows to be a tolerably definite chemical composed of nucleic acid (a complex organic acid rich in phosphorus) and albumin. And thus we reach the remarkable conclusion that inheritance may, perhaps, be effected by the physical transmission of a particular chemical compound from parent to offspring.[12]

Sixty years later, Wilson's conjecture had become a central theory of the transmission mechanism of heredity.

The recognition of the nucleus as the physical vehicle of inheritance, Wilson pointed out in 1900, "is now widely accepted, but acceptance requires rejection of the theory of germinal localization" (*i.e.*, that adult characters, while not preformed in the germ, are prelocalized in the egg cytoplasm). As Wilson points out,

> de Vries in 1889 brought this conception into relation with the theory of nuclear idioplasm by assuming pangens in the nucleus, migrating into the cytoplasm step by step during ontogeny. Pangens are not all germs as in Darwin's theory but ultimate protoplasmic units of which cells are built and which are bearers of hereditary qualities.

Thus one might say that a chromosome theory of heredity existed, as a speculative construct, before a theory of the gene. But, except for its indirect influence in leading de Vries into experimental plant breeding, it did not by itself lead to the discovery of the gene.

---

[12] I am grateful to my colleague, Professor John A. Moore, who called to my attention this passage in Wilson's *An Atlas of the Fertilization and Karyokinesis of the Ovum* (1895).

– Chapter 5 –

DISCONTINUOUS VARIATION: 1753–1900

Heredity and variation, the concepts of continuity between generations and of the differences which appear within that continuity, have always to be considered together. Once it was the fixity of species, the maintenance of type, and the occasional departures from type which required their association. Later, when the particulate nature of heredity became known, it was realized that only those particles or genes could be recognized which had varied or mutated to produce alternative forms. What we know about heredity, we know because of the existence of hereditary variations.

The Mendelian view of heredity and the theory of mutation were first promulgated in the same year and, as it happened, by the same man, de Vries. He reached both ideas, that of pangenes and that of mutation, by the same route, and they were closely connected in his mind. In the second volume of his *Mutationstheorie* (1903, p. 643), (as translated by Farmer and Darbishire) he has this to say:

> As a hypothesis, pangenesis serves a heuristic object; as a theory, it must serve as a basis from which a deeper insight into the nature of the living substance may be obtained. I have not much to say here as to its heuristic

value, since for myself pangenesis has always been the starting point of my enquiries; at first only in a theoretical way, but afterwards also for the experimental investigations described in this book.

Especially is it this hypothesis which has led me to search for mutations in the field, because I hoped in this way to find facts which would throw a more immediate light on the bearers of the hereditary characters, and thereby on the theory of heredity in general.

Then follows a very revealing footnote:

I should like to insert here the following little coincidence. My *Intracellular Pangenesis* was written during the summer holidays, spent near Hilversum in 1888, and the often described locality of *Oenothera Lamarckiana* was only about ten minutes walk away.

The connection between units of inheritance and variation had been recognized by Darwin, and de Vries was impressed by this, for he paraphrases (p. 644) from the 1876 edition of Darwin's *Variation of Animals and Plants under Domestication* the passage of p. 390 of Volume II in which Darwin points out how his hypothesis of pangenesis (which, as we have seen, was quite different from de Vries') calls for "two distinct groups of causes" of variation, one due to changes in number or to rearrangement of gemmules which are themselves not modified, the other to changes in the particles themselves. The first category in de Vries' theory provided the basis for fluctuating variability, the second for progressive mutability. These distinctions sound like two of the categories recognized later—those of quantitative or polygenic variation on the one hand and "major" gene mutations on the other; but the resemblance is superficial since the units involved have different properties. Again, it is the heuristic value of thinking in terms of units which is emphasized.

The association between hereditary units and variation goes back further than Darwin. Bentley Glass (1947) cites an interesting passage from Maupertuis' *Système de la Nature* (1751):

> Could not one explain by that means [mutation] how from two individuals alone the multiplication of the most dissimilar species could have followed? They could have owed their first origination only to certain fortuitous productions, in which the elementary particles failed to retain the order they possessed in the father and mother animals; each degree of error would have produced a new species; and by reason of repeated deviations would have arrived at the infinite diversity of animals we see today.

As we see time and again in the history of genetics, as of other sciences, similar ideas recur; and it is less often novelty or originality in the imagined explanations than forms of proof or new designs for testing hypotheses which influence the direction of growth of a science.

Thus, a span of years elapsed between the first published record of the origin of a new variety by mutation, *viz.*, Sprenger's discovery of the cut-leafed form of *Chelidonium majus* in 1590, and de Vries' discovery of the first mutants in *Oenothera Lamarckiana* about 300 years later, in 1886. In the interval there were repeated and increasingly frequent observations of new varieties, both of plants and animals, which appeared suddenly and thereafter reproduced the new form. Darwin brought many of these facts together and was the first to provide a general theory for explaining the relation of these to evolution, not attaching much importance to major discontinuous changes or sports as material for evolution because of their disturbing effect on the internal equilibrium and adjustment to the environment. Stubbe provided, in 1938, a history of the occurrences of sudden variation and later reviewed it and brought it up to date (1963). But advancement of genetics as such did not result from any of the scattered instances of discontinuous variation and probably could not have arisen from Darwin's efforts until de

Vries made the essential synthesis in his *Mutationstheorie* (1901–1903).

The trend in this direction was clearly marked by the major work of William Bateson, *Materials for the Study of Variation*, which appeared in 1894. This was a bold and original attack on that part of the theory of natural selection which assumed that it operated primarily on small continuous variations. The subtitle of the book revealed its chief thesis "treated with especial regard to DISCONTINUITY in the origin of species" (the capital letters are Bateson's). Bateson's conclusion was that the discontinuity of species resulted from the discontinuity of variation, and detailed data on discontinuous variation filled most of the nearly 600 pages of that book. Even earlier, Galton (1889) had recognized the importance of discontinuous variation as a source of the variety on which natural selection operates. Following Bateson, the Russian botanist Korschinsky, in 1900, elaborated on the occurrence of similar forms of variation to which von Kölliker in 1864 had given the name "heterogenesis." (For an account of Korschinsky's work and influence, see Stubbe, 1963). Already, in 1899, Korschinsky had concluded that among garden plants all new characters originated by heterogenesis and that such mutations were of much more frequent occurrence than had previously been supposed and provided a more important source of evolutionary material than the natural selection of small variations.

However, it is to de Vries that we owe the introduction of mutation theory into genetics. It was his detection of mutations in planned experiments and his testing of the mutants that was responsible for this, and not the proof of discontinuity in the origin of species. His work dealt actually with "experimental evolution" and gave for the first time the hope of directly observing and possibly controlling evolutionary processes.

I have chosen to refer to the period of genetics from 1900 to 1910 as the "period of Mendelism." Perhaps it would be better to think of it as "Mendelism and mutation," for the second concept had an influence on genetics almost as great as the first, although of course dependent upon the first. In some ways the publication of the first volume of de Vries' great work in 1901 made a greater impression on biology than the rediscovery of Mendel's principles. It certainly aroused greater opposition. The mutation theory dealt with what appeared to be a more general problem, the mechanism of evolution, while no such claims of universal applicability were at first made for Mendelism. Then, too, the new facts revealed by de Vries' long-continued observation of thousands of plants of Oenothera through many generations, and the regular occurrence within his cultures of striking and discontinuous variant types—mutants—which bred true from the beginning, were spectacular and dramatically presented. His claim was to have discovered a mode, perhaps *the* mode, by which new species originated, for he called each of the variant types of Oenothera that appeared an elementary species. He supposed the new complex of characters to have arisen by a single step which he referred to as mutation. It was this claim, in direct opposition to that of selection among many small variations, as assumed by Darwin's successors, which aroused both strong opposition from the selectionists and support from some experimentalists who repeated and extended de Vries' breeding experiments. The beneficial effects of this controversy on both theoretical and practical genetics were striking. Coming at the same time as the recognition of the gene—or, as some called it, the "unit character"—a view of the origin of new hereditary characters by single steps focused attention on an essential question of both genetics and evolution. Most important of all it showed that such questions could be studied experimentally.

The story of de Vries' work on mutation has often been told and will be found in most textbooks of classical genetics. In 1886 he found, in an abandoned potato field near Amsterdam, large numbers of plants of the evening primrose, *Oenothera Lamarckiana*, growing wild. Since this is an American species it was assumed to have escaped from the European gardens where it was cultivated. Among the wild plants were two new varieties each of which differed from the normal form of the species in a number of characteristics. De Vries' first reference to these in his *Mutationstheorie* (1901) indicates their important features:

> Both *Oenothera brevistylis* and *Oenothera laevifolia* come perfectly true from seeds as will be shown later on. They differ from *Oenothera Lamarckiana* in numerous characters, and are therefore to be considered as true elementary species.

This experience repeated itself with descendants of the wild plants of *Oenothera Lamarckiana* which de Vries bred in large numbers in his experimental garden. In a total population of some 50,000 plants raised, about 800 were mutants of seven different types to which he gave specific names. Others soon confirmed de Vries' claim that this plant was in a mutable phase by obtaining the mutants which de Vries had observed plus many others both in *Oenothera Lamarckiana* and in some other species of Oenothera.

As far as genetics is concerned, it is probably of less importance that de Vries' two main claims were not sustained by later work than that a new and profitable direction of study had been initiated. His claim to have discovered a mode by which new elementary species originated in a single step and his claim that all Oenothera mutants were examples of one process of mutation were both disproved by the discovery that the original mutants were due to several different processes. Some were due to rare occurrences of recombination in heterozygotes maintained in a balanced system by the lethality of homozygotes; others were due to changes in whole chromosome sets (polyploidy), such as doubling to produce a tetraploid, like his *gigas* species, or a haploid like *nanella*, the dwarf. Only two proved to be due to the origin by mutation of a new Mendelian allele, the process to which the term mutation is now generally restricted. The "mutation theory" of the origin of new variations was thus founded on occurrences most of which were not new; and the varieties originating from these occurrences were not "species" but a heterogeneous collection of forms, none of which could take rank as a new taxonomic category. In the process of investigating de Vries' claims, however, gene mutation was recognized as the essential step in the origination of hereditary variety, and the direction was discovered in which the relation between the behavior of genes and of chromosomes and the breeding behavior of complexes of phenotypic characters could be explored.

I am reminded by such occurrences that the road to the solution of a scientific puzzle such as that presented by the Oenothera "mutations" may resemble the way to the solution of a double acrostic or crossword puzzle. It seems that sometimes the important thing is simply to keep on writing in letters (steps in the solution) even when they are wrong and to arrive at a correct solution by serial correction of earlier errors. This means that, in the end, it is more profitable to have errors to correct than a blank page.

Much of the subsequent history of the mutation concept belongs to a later period and will be reviewed in Chapter 19.

– Chapter 6 –

WILLIAM BATESON AND
THE BIRTH OF GENETICS

AMONG THE CREATORS of modern genetics the English zoologist William Bateson (1861–1926) occupies a place comparable to that of de Vries and Correns. He came to the subject prepared by long thinking on discontinuous variations as forming the materials for evolution. He had been influenced in this direction, he has told us, by Professor W. K. Brooks of Johns Hopkins University, with whom he spent the summers of 1883 and 1884 at the Hopkins summer laboratory at Hampton, Virginia. Brooks was then writing a book on heredity, and his notion "that there was a special physiology of heredity capable of independent study, came as a new idea" to Bateson. Bateson's doubts about the adequacy of the theory of continuous variation led him on long tours of observation in the field: eighteen months in Turkestan, then a shorter period in Egypt. After his tours, Bateson settled into a research studentship at Cambridge University, and, in 1894, he produced his great compilation, *Materials for the Study of Variation*.[13]

Before 1890, he had begun work on experimental hybrid-

[13] All of Bateson's papers on genetics are in the collection edited by R. C. Punnett (*cf.* Bateson, W., 1928); biography in Crowther, 1952.

ization with cultivated plants. This early breeding work was undertaken to determine how discontinuous variations were inherited and to provide material for his controversy with his close friend W. F. R. Weldon, an Oxford zoologist. Weldon considered continuous variation the important source of evolutionary change and under the influence of Galton devoted himself to biometric study in support of his thesis. In 1895, Bateson's first open break with Weldon occurred, and as the breach widened the two former friends became, after 1900, leaders of two warring camps, the biometricians and the Mendelians.

Bateson's plant-breeding experiments were carried out wherever opportunity afforded until, in 1897, a government grant to the Evolution Committee of the Royal Society enabled him to begin organized and continuous work both with animals and plants at the Cambridge Botanic Garden. This brought him into association with the Royal Horticultural Society at whose meeting on July 11, 1899, he presented a paper, "Hybridization and Cross Breeding as a Method of Scientific Investigation." This paper shows clearly how close he had come to the methods of study, the statement of the problem and statistical analysis of the offspring of hybrids which had led to Mendel's success. However, his breeding experiments with plants and with various forms of comb in poultry had not yet yielded sufficient data to permit interpretation.

In her memoir of her husband, Beatrice Bateson has recorded that on May 8, 1900, Bateson took the train to London to deliver a second paper to the Royal Horticultural Society, "Problems of Heredity as a Subject for Horticultural Investigation." Mrs. Bateson writes:

On his way to town to deliver it he read Mendel's actual paper on peas for the first time. As a lecturer he was always cautious, suggesting rather than affirming his own convictions. So ready was he however for the simple Mendelian law that he at once incorporated it into his lecture.

From that day on, Bateson devoted all his enthusiasm and literary gifts, which were considerable, to promulgating what he soon came to call "Mendelism." He immediately had Mendel's paper of 1866 translated and published with footnotes in the *Journal of the Royal Horticultural Society* (1900). It was this translation which brought Mendel's paper to the attention of the English-speaking world. Bateson became the most ardent advocate of the new view of heredity, feeling himself a self-appointed *vox clamantis in deserto* to an extent which we find difficult to understand today until we learn of the opposition of systematists, conservative zoologists and botanists, and the especially vindictive attacks on Mendel by Weldon and the biometricians.

It was one such attack on Mendelism by Weldon in the first volume of *Biometrika* (1902) that called forth Bateson's *Mendel's Principles of Heredity, A Defense* (1902). This became the first textbook presentation of the elementary facts of genetics. The tone of the whole brief book is indicated by the first sentence: "In the study of evolution progress had well-nigh stopped." Then came the great break-through, Mendel's principles, which, however, some "established prophets" failed to understand. Bateson says his purpose is to expound Mendelism and defend it from Weldon's attacks "by no more elaborate process than a reference to the original records."

Then referring to the source and cause of the attack, Bateson continued (1902, preface, p. iii):

Mr. Galton suggested that the new scientific firm [of editors for *Biometrika*] should have a mathematician and a biologist as partners; and—soundest advice—a logician retained as consultant. Biologist surely must one partner be, but it will never do to have him sleeping. In many well-regulated occupations there are persons known as "knockers-up" whose thankless task it is to rouse others from their slumbers and tell them work time is come round again. That part I am venturing to play this morning and if I have knocked a trifle loud it is because there is need.

Weldon's paper, "Mendel's Laws of Alternative Inheritance in Peas" (*Biometrika* 1, Pt. II), reached Bateson on February 8, 1902, and he seems to have written his "defense" immediately. Weldon's conclusion that "Mendel's results do not justify any general statements concerning inheritance in cross-bred peas" is answered by Bateson:

> After close study of his article it is evident to me that Professor Weldon's criticism is baseless and for the most part irrelevant and I am strong in the conviction that the cause that will sustain damage in this debate is not that of Mendel.

Bateson then describes Mendel's work, his own repetition of it, and work of other hybridizers which could now be interpreted by Mendel's explanation. How thoroughly Bateson understood the new principles of segregation is shown by his statement: "His experiments are worthy to rank with those which laid the foundation of the Atomic Laws of Chemistry." Mendel's principles are "lying at the very root of all conceptions not merely of the physiology of reproduction and heredity but even of the essential nature of living organisms." This vigorous championship did not itself add to knowledge, of course, but it was one of the factors in the rapid spread of knowledge (if not always of acceptance) of the new views and it had a great influence in England and especially in the United States. W. E. Castle, a chief pioneer of genetics in the United States, said (1951) of Bateson: "He was the real founder of the science of genetics as well as the one who gave it that name."

Important though Bateson was as advocate, his own work as an experimental breeder helped to lay good foundations for the analyses of the transmission mechanism. The five "Reports to the Evolution Committee" which appeared between 1902 and 1910 contain the detailed results of breeding experiments by Bateson and his collaborators, Miss E. R. Saunders, and, later, R. C. Punnett, C. C. Hurst, Florence M.

Durham, L. Doncaster, and others. Mendel's rules were confirmed and extended in a number of different plant species, and the first "Mendelian" characters in animals (poultry) were reported. At about the same time, Lucien Cuénot (1902) showed that certain coat-color differences in mice conformed to Mendel's rules. Thus, Bateson and Cuénot were the first to demonstrate experimentally the extension of Mendel's theory to animals.

In Bateson's introduction to Report No. 1 (1902, p. 12) we find the clear recognition of the essence of Mendelism:

> The essential part of the discovery is the evidence that *the germ cells or gametes produced by crossbred organisms in respect of given characters may be of the pure parental types and consequently incapable of transmitting the opposite character*: that when such pure similar gametes of opposite sexes are united together in fertilization, the individuals so formed and their posterity are free from all taint of the cross; *that there may be, in short, perfect or almost perfect discontinuity between these germs in respect of one of each pair of opposite characters.*

At the end of this report (pp. 126–130) Bateson suggested the terminology which the new field quickly adopted.

> By crossing two forms exhibiting antagonistic characters, cross-breds were produced. The generative cells of these cross-breds were shown to be of two kinds, each being pure in respect of *one* of the parental characters. This purity of the germ cells, and their inability to transmit both of the antagonistic characters, is the central fact proved by Mendel's work. We thus reach the conception of unit-characters existing in antagonistic pairs. Such characters we propose to call *allelomorphs*,[14] and the zygote formed by the union of a pair of opposite allelomorphic gametes, we shall call a *heterozygote*. Similarly, the zygote formed by the union of gametes having similar allelomorphs, may be spoken of as a *homozygote*. Upon a wide survey, we now recognize that this first principle has an extensive application in nature. We cannot as yet determine the limits of its applicability, and it is possible

[14] Later shortened to *allele*, which is now in general use.

that many characters may really be allelomorphic, which we now suppose to be "transmissible" in any degree or intensity. On the other hand, it is equally possible that characters found to be allelomorphic in some cases may prove to be non-allelomorphic in others.

This leads to a point of great importance to the evolutionist. We have been in the habit of speaking of a variation as discontinuous, in proportion as between it and other forms of the species intermediates are comparatively scarce when all breed freely together. In all cases of allelomorphic characters we can now give a more precise meaning to this description. It must now be recognized that such a population consists, in respect of each pair of allelomorphs of three[15] kinds of individuals, namely, homozygotes containing one allelomorph, homozygotes containing the other allelomorph, and heterozygotes compounded of both. The first two will thus always form discontinuous groups, and the degree to which the heterozygotes form a connecting group will depend on whether one allelomorph regularly or chiefly dominates in the heterozygotes, or the allelomorphic characters completely or partially blend in the heterozygote. *Such discontinuity will in fact primarily depend not on the blending or nonblending of the characters, as hitherto generally assumed, but on the permanent discontinuity or purity of the unfertilized germ-cells.*

It is to be noticed that Bateson did not at first use the term "segregation," the name we now give to Mendel's first principle. That term came from de Vries, who used *"Spaltung"* (splitting) in German. For Bateson the same idea was expressed by "purity of the gametes"; later (p. 150) by segregation.

Bateson's prescience is revealed in a footnote to Report No. 1 (p. 133) dated December 17, 1901. He had discussed the possibility of the persistence of hidden recessives without homozygotes ever having been seen. The footnote is as follows:

In illustration of such a phenomenon we may perhaps venture to refer to the extraordinarily interesting evidence lately collected by Garrod regarding the rare con-

[15] Four, if reciprocal heterozygotes are not identical.

dition known as "Alkaptonuria." In such persons the substance, alkapton, forms a regular constituent of the urine, giving it a deep brown colour which becomes black on exposure.

The condition is extremely rare, and, though met with in several members of the same families, has only once been known to be directly transmitted from parent to offspring. Recently, however, Garrod has noticed that no fewer than five families containing alkaptonuric members, more than a quarter of the recorded cases, are the offspring of unions of *first cousins*. In only two other families is the parentage known, one of these being the case in which the father was alkaptonuric. In the other case the parents were *not* related. Now there may be other accounts possible, but we note that the mating of first cousins gives exactly the conditions most likely to enable a rare and usually recessive character to show itself. If the bearer of such a gamete mate with individuals not bearing it, the character would hardly ever be seen; but first cousins will frequently be bearers of *similar* gametes, which may in such unions meet each other, and thus lead to the manifestation of the peculiar recessive characters in the zygote.[16]

It was Bateson who helped to put Garrod on the right track in human biochemical genetics.

The Third Conference on Hybridization and Plant Breeding was convened by the Royal Horticultural Society in 1906, but by the time its proceedings were published in 1907 it had become a conference on genetics. The change was due to the following passage in Bateson's Inaugural Address to the Conference on July 31, 1906 (1907, p. 91):

> I suggest for the consideration of this congress the term Genetics, which sufficiently indicates that our labours are devoted to the elucidation of the phenomena of heredity and variation: in other words to the physiology of descent, with implied bearing on the theoretical problems of the evolutionist and the systematist, and application to the practical problems of the breeder, whether of animals or of plants.

[16] See Garrod, 1899, p. 367, and 1901.

The name "genetics" appeared first in print in 1906 in a review by Bateson of a book by J. P. Lotsy (*cf.* W. Bateson, 1928, p. 442):

> As the moment is favourable, may it be suggested that the branch of science the rapid growth of which forms the occasion of Professor Lotsy's book should now receive a distinctive name . . . To avoid further periphrasis, then, let us say genetics.

The word had first been used by Bateson in a letter of April 4, 1905, to his friend Prof. Adam Sedgwick (*cf.* B. Bateson, 1928, p. 93):

> If the Quick Fund were used for the foundation of a professorship relating to Heredity and Variation the best title would, I think, be "The Quick Professorship of the study of Heredity." No single word in common use quite gives this meaning. Such a word is badly wanted and if it were desirable to coin one, "Genetics" might do. Either expression clearly includes Variation and the cognate phenomena.

Bateson had hoped that, if established, this professorhip might come to him; but in this as in previous applications he was disappointed. He maintained himself and often his plants and animals on a meager income from studentships and a fellowship of his college, St. John's, Cambridge, which for many years he served as steward. He finally was appointed to a professorship of Biology in 1908 at the age of forty-seven; but even then the post had been created only for a five-year period from a special gift.

Bateson's 1908 inaugural lecture, "The Methods and Scope of Genetics" is one of the most interesting of all of his works and shows at its best his ability as a writer of clear, pungent prose. It was in this lecture that occurred his most often quoted remark (1908, p. 22):

> Treasure your exceptions! When there are none, the work gets so dull that no one cares to carry it further. Keep

them always uncovered and in sight. Exceptions are like the rough brickwork of a growing building which tells that there is more to come and shows where the next construction is to be.

Bateson occupied this professorship only until 1910, when he left Cambridge to become director of the John Innes Horticultural Institution at Merton. Here he assembled a group of scientific collaborators and led them in an enterprise which clearly had animated much of his earlier work, that of bringing to the practical breeder the knowledge and methods of science and of gaining from gardeners and farmers knowledge of new problems. This probably compensated Bateson to some extent for the lack of a "school" at Cambridge, for there had been collaborators and assistants but seldom a body of students to develop the field he had founded.

Bateson's own discoveries in genetics grew out of his experimental investigations of apparent exceptions to Mendel's rules. It was Bateson's skepticism about conclusions resting on anything except detailed breeding analysis which, as J. B. S. Haldane has said, prevented Mendelism from becoming a dogma. Bateson was a hard man to convince and to keep convinced. As Haldane stated in his 1928 essay (p. 144): "He had many disciples but was never himself of their number." Bateson's critical attitude extended equally to himself and to others. It was this which led him to the explanation of the departures from the ratios of different characters expected on Mendelian principles. The first such explanation was that two or more different Mendelian units ("unit characters" as they were then known) could interact to produce an effect different from that of either unit, as when two races of fowls, each owing its white plumage to a different mutant gene, could produce, when crossed, only colored fowls. This kind of complementary interaction meant that at least two different genes were concerned with pigment synthesis and that they could mutate separately and be inherited independently of each other. The suppression of effects of one gene by those of a different one (epistasis) or the failure of a white fowl, for example, to show the effects of a dominant gene for color (hypostasis) were first worked out by Bateson and Punnett and served as models for many other breeding analyses in animals and plants. At about the same time Cuénot (1903) showed that albino mice could transmit different genes affecting coat colors, without showing the effect of these.

Bateson with his colleague Punnett discovered linkage experimentally, although de Vries and Correns had both foreseen the groupwise inheritance of separable elements, and Sutton had predicted it. Bateson and Punnett called the phenomenon "gametic coupling" and recognized its importance but not its relation to chromosomes. The explanation they proposed was called "reduplication" and it assumed that the characters which tended to stay together in inheritance did so because gametes containing them underwent additional post-meiotic divisions. This assumption was soon replaced by Morgan's hypothesis that coupled genes were in the same chromosome and were subject to separation or recombination by crossing over; but Bateson resisted this generalized interpretation for eight years after evidence for it had convinced most geneticists. So, too, did he cling to his hypothesis that dominant genes represented the presence of something that was absent in the corresponding recessive allele. The presence and absence hypothesis also failed to survive the early years of genetics.

As soon as the first success of the chromosome theory became apparent, about 1912, Bateson, as a geneticist, seems to have been left behind. Although, in 1922, he handsomely acknowledged the achievements of Morgan's school, he seems never to have been really convinced by them, and in his last paper (1926) returned to his former skepticism.[17] Bateson's

[17] Bateson's colleague, R. C. Punnett, revealed in a charming set of reminiscences (1950) the reason why he and Bateson . . . "managed to miss the tie-up of linkage phenomena with the chromosomes. The answer is Boveri. We were deeply impressed by his paper (On the Individuality of the Chromosomes) and felt that any tampering with them by way of breakage and recombination was forbidden." This,

last investigations were concerned with segregation and he dealt extensively with variegation, bud sports, chimaeras, root cuttings, and similar phenomena in plants which he thought could provide evidence that segregation was not restricted to the meiotic divisions. Here again he failed to convince geneticists generally.

Bateson's great contributions were made in his earlier years. His criticism of the theory of natural selection as acting on continuous variations, while itself ill-founded, stimulated him to search for proof of discontinuity. This search led him to experimental breeding and to founding classical genetics as a recognized biological discipline.

---

as Stern (1950) quickly pointed out, was due to a misreading of Boveri, who, in 1904, had discussed the pairing and exchange of parts of homologous chromosomes and the breeding evidence required to demonstrate it.

– Chapter 7 –

THE CONTRIBUTIONS OF
CORRENS AND TSCHERMAK

THE PART that Carl Correns played in establishing
genetics on a sound basis was quite different from that of
Bateson, the importunate advocate. Correns was a quiet, reserved research scientist devoted only to his work of convert-
ing at least one sector of biology into an exact science. He
was largely self-taught in botany, his doctoral dissertation,
written in 1889, on the growth of membranes in algae having
been done at home with his own means. His professor was
Carl von Naegeli, who was no longer lecturing. From Naegeli,
Correns first learned of Mendel's Hieracium crosses. Mendel's
work with peas and the principles derived from it, however,
were not mentioned by Naegeli.

Correns began his own plant-breeding work while privat-
dozent in Tübingen, 1892–1902. It was here that he discovered experimentally the explanation of the phenomenon of
xenia—the direct influence of foreign pollen on the endosperm
of the seed. The occurrence of xenia in maize had long been
known (cf. Zirkle, 1935). Correns' interpretation of xenia
confirmed S. Nawaschin and Guignard's cytological discovery
of the triploid nature of the endosperm, one set of chromosomes being derived from a second haploid nucleus of the

pollen grain which in the case of xenia contributes a dominant gene, expressed in the endosperm. Pursuing this question and others relating to the functioning of pollen grains, Correns carried out many hybridizations with stocks, maize, beans, peas and lilies. In the results with maize and peas he noticed the occurrence of sharp segregation of alternative characters, and this led him, in 1899, to Mendel's 1866 paper. Correns immediately saw that the explanation he had just reached had been worked out in masterly fashion thirty-three years earlier. What Correns stated in another connection is pertinent here: "Whoever supports with his arguments the establishment of a correct view of another person, may have accomplished more than he who first proposed it." Thus it was that claims to priority and the struggle for credit and recognition, so frequent among research scientists, seem not to have troubled Correns. This feeling accounts also for his indignation when, in de Vries' preliminary summary of his findings, Mendel's name was not mentioned (*cf.* page 3).

Correns' own works, published chiefly in the *Proceedings of the Prussian Academy of Sciences*, were in their own period not well known. They reveal that he recognized both coupling of Mendelian characters and multiple effects of single genes. He gave the first proof of the sex-difference as a Mendelian character, and was the first to anticipate the view proved many years later that sex-determination is not fixed but that each part of each gamete contains the developmental potentiality of the opposite sex. He obtained the first conclusive proofs of non-Mendelian (cytoplasmic) inheritance. His last paper (1937) was a masterly review of evidence for non-chromosomal genes. As the leading geneticist of Germany (although he was not as well known in other countries) he became, in 1913, the first director of the newly founded Kaiser Wilhelm Institut für Biologie in Berlin-Dahlem. Here for the first time he had the means to carry on extensive research. Unfortunately, most of the records of his work were still unpublished when, during the bombing of Berlin in 1945, all of them were destroyed. Correns' influence was felt in all fields of genetics with the possible exception of population genetics in which he did not express special interest.

Of the three rediscoverers there can be little doubt that it was Correns who had the broadest view of the problems and principles of genetics. He alone of the three observed, interpreted and emphasized the recombination of independent pairs of characters as an integral part of "Mendel's Law." It was Correns who first identified in his first report (1900) the segregation of alleles with the reduction division in meiosis. He foresaw that dominance would not prove to be an essential part of Mendel's Law and that other departures from Mendel's basic rules could lead to deeper understanding of the transmission mechanism of heredity, as in fact happened later with linked characters.

The third rediscoverer of the principle of segregation, Erich von Tschermak-Seysenegg, took his doctor's degree in botany at the University of Halle in 1895 and then for two years worked as a volunteer in horticultural and seed-breeding establishments in Germany. He returned to Vienna, where his father was professor of mineralogy, in January, 1898, and in a paper which he presented at a meeting of the Agriculture and Forestry Club he spoke of his newly aroused interest in plant breeding. While waiting for an assistant's place to open for him at the Agricultural College in Vienna, he went, early in 1898, to work under Professor McCleod, director of the Botanical Garden in Ghent. There, at McCleod's suggestion, he began experiments to extend Darwin's observations on the effects of self- and cross-fertilization, and he chose to experiment with races of peas which could be quickly reared in the greenhouse. He crossed varieties differing in some of the same characters which Mendel had used (tall with dwarf, yellow-seeded with green, round-seeded with wrinkled) and in July 1898 took the crossbred seeds with him when he returned to Vienna. There in a small experiment station at Esslingen he bred and observed F_2 progeny and began other crosses with the primary object of studying the immediate effect of crossing, *i.e.*, xenia. When he came to work up his results in the fall of 1899, he noticed the occurrence of dominance. He wrote in a letter dated January 7, 1925 (*cf.* Roberts, 1929, p. 345) :

> In counting out the seed-characters, the ever-recurring number relationship of 3 : 1 naturally could not escape me, any more than the number relation of 1 : 1 on backcrossing of green-seeded peas with hybrid pollen of the F_1 generation.

While working on his results, in the autumn of 1899, he found a reference to Mendel's paper in Focke's *Pflanzenmischlinge* and on the same day got and read it. He then prepared the account of his own rather limited observations as a dissertation for obtaining the grade of dozent. After this had been submitted, he was surprised to find similar facts and interpretations in the two papers of de Vries and one of Correns. He prevailed upon the editor of the *Zeitschrift für das Landwirtschaftliche Versuchswesen im Oesterreich* to publish in advance the "separates," or preprints of his paper which thus appeared in May, 1900, even before the abstract sent June 2 to the *Berichte der deutschen botanischen Gesellschaft.*

Tschermak's results pertained primarily to the questions raised by Darwin concerning effects of crossing in increasing height and other aspects of vigor; but the discovery of Mendel's explanation of dominance put all this in a rather different light. Tschermak noted some indication of incomplete dominance and of greater influence of the female parent on certain seed characters, neither of which had been noted by Mendel. On the whole, however, Tschermak's paper can be said to represent a confirmation of Mendel's results for single pairs of characters rather than a rediscovery of Mendel's principles, since Tschermak did not develop the theories of segregation or independent assortment—although there is no doubt that he later came to recognize their importance.[18] In

[18] Tschermak has published several accounts of his part in the establishment of genetics (1951, 1956, 1960), and a full account including letters from Tschermak is given by Roberts (1929).

the same year (1900), Tschermak arranged to have Mendel's original paper republished in Wilhelm Ostwald's *Klassiker der exakten Wissenschaften,* although before that occurred Karl von Goebel had already reprinted all three of Mendel's papers in *Flora* in 1901.

Tschermak's influence was exerted primarily through plant-breeding practices and his visit to the Swedish Plant Breeding Institute at Svalöf in 1901 probably affected the later work of that very important center of crop improvement.

Part II

A DECADE OF MENDELISM: 1900–1909

– Chapter 8 –

TESTING MENDEL'S RULES

LET US SEE what had been added to Mendel's principles by the end of the first decade after 1900, say up to the time of publication of Johannsen's *Elemente der exakten Erblichkeitslehre* in 1909. A great deal of stock-taking went on in 1909, since that year marked both the 100th anniversary of Darwin's birth and the fiftieth year of the publication of the *Origin of Species.*

Of the three elementary rules first discovered by Mendel, one (dominance) had been shown by Correns and many others to be no rule at all. The fact that in each of seven pairs of "differentiating characters" for which Mendel reported statistical data one member of the pair appeared to be dominant over the other proved to be due in part to coincidence and in part to superficial examination. For example, more careful examination of some heterozygotes by the microscope revealed effects of the recessive gene, and this was found to occur more commonly than complete dominance.

The principle of independent assortment, or recombination as limited to genes in different chromosome pairs, had not yet been clarified, although Sutton's interpretation had made it evident that proof of this limitation was to be anticipated. Coupling of characters had been observed by Correns but had received no convincing interpretation. A. D. Darbishire (1904) published data which was later shown to prove linkage between two coat-color genes in the house mouse. Linkage of pairs of characters in parental associations had been discovered by Bateson and Punnett, but the interpretation proved to be wrong.

The principle of segregation had withstood all assaults and had emerged as the cornerstone of transmission theory. "Mendelism," as this body of principles came to be called, had been strengthened by the critical examination of apparent exceptions to the rules stated by Mendel. Many of the exceptions were found to be due to peculiarities in the expression of genes in combination such as complementary interaction and epistasis. The principle even survived the first case of selective fertilization discovered by Correns (1902), in which an exception to the random union of gametes assumed by Mendel was explained by effects of a gene on the functioning of pollen grains (*cf.* p. 102).

Just at the end of this period appeared the clue which led to the exclusion of the possibility that contamination of gametes in heterozygotes could explain apparent blending inheritance. This finding was the result of H. Nilsson-Ehle's demonstration of recombination of multiple genes affecting quantitative characters. The theory of gene change by mutation encompassed originally only a pair of alternatives, but cases that could only be explained by the existence of multiple alleles were encountered by Cuénot (1905).

An orderly terminology had been devised and at the end of the period the terms "gene," "genotype," and "phenotype" had been introduced, followed by the useful concept of the "norm of reaction" as the expression of the genotype.

It was fairly well settled that new hereditary variants arose by a discontinuous process known as mutation, but the nature of the process was unknown. The proof that the variability found within pure lines—homozygotes—was not inherited since such lines could not be changed by selection, disposed of environment as responsible for hereditary change and, together with Mendelism, excluded inheritance of acquired characters as a source of new hereditary variants. The lack of influence of the phenotype on the genotype (the concepts introduced by Johannsen in 1909) was soon shown in another way by the ovarian transplantation experiments of W. E. Castle and John C. Phillips.

The groundwork had been laid for the interpretation of sex differences in Mendelian terms, and the inheritance of the first sex-linked traits (in the moth Abraxas) had been worked out. The sex-linked inheritance of the barred plumage pattern of Plymouth Rock fowls had been well established by H. D. Goodale, W. J. Spillman, and others in 1909, who had shown in this case as in that of the moth, that the female was the digametic sex. These cases had been worked out and correctly interpreted before the establishment of the first case of sex-linkage (of the opposite type) in Drosophila (Morgan, 1910); but it was the latter which revealed the correlation between the inheritance of sex chromosomes and of sex-linked traits.

Sex chromosomes had been observed by H. Henking in 1891, and their function was pointed out soon after the re-discovery of Mendel's principles. Qualitative differences among other pairs were shown by Boveri to be highly prob-able. If the earlier period (1880–1900) had been marked by "preparing cytology for Mendelism," the period 1900–1909 was one of interpreting chromosome structure and behavior for the purposes of cytology rather than of genetics. Just at the end of the period, however, came the observation by F. A. Janssens (1909) of chiasmata during synapsis which became the cytological counterpart of Morgan's theory of crossing over.

The first steps in physiological genetics—toward a theory of "gene action"—were taken early in this period. Garrod, the English physician and biochemist, aided by Bateson as ge-netical advisor, was responsible for initiating, in 1902, research which many years later led to the development of human biochemical genetics. Garrod's great contribution was the idea that each "inborn error of metabolism" (he dealt with four such diseases in his Croonian lectures of 1908) was due to a block or interruption at some point in a metabolic reaction sequence and that the block was due to the congenital deficiency of a specific enzyme. This idea, and Garrod's work, had little influence on the development of genetics at that time. However, when the same idea was reinvoked in the late 1930s to explain the manner of genic control of reaction sequences leading to eye pigment production in Drosophila, it led to a generalized metabolic genetics and was profitably exploited even before specific enzymes controlling particular steps had been identified.

The first proof of Mendelian heredity in man had come from Garrod's work on alkaptonuria (1902). But even when he extended the interpretation to other diseases (1909), the response of biologists was meager, and little serious work on human heredity was done. Further, research on man seems to have been handicapped because of interest in the immediate social applications of genetics, such as eugenics, and no important addition to principles, apart from Garrod's theory, resulted (Dunn, 1962).

Population genetics, which began in theoretical form with Wilhelm Weinberg in 1908–1909, made little impression then on those who might have been expected to be interested in human heredity. Papers by Castle (1903), Karl Pearson (1904) and by G. H. Hardy (1908) showed that the idea of genetical equilibrium was in the air. It was, in fact, implicit in Mendel's principles as soon as these were thought of in cross-fertilizing, random-breeding populations. Weinberg's work, however, did not get much attention until, in the 1930s, genetical theories of evolutionary processes began to be developed in earnest.

The condition of stasis which evolutionary thinking in genetics had reached in this decade was not shared by the field from which the methods developed by which evolutionary change was to be studied. Biometry, founded by Galton in the 1880's, was vigorously developed during this period by Pearson. But the failure of both men to appreciate the essential features of Mendelian heredity led their school into opposition to Mendelism and hindered rather than helped the testing and extension of Mendel's principles. However, Johannsen, Nilsson-Ehle, and others interested in using statistical methods in genetics applied the methods of quantitative variation and inheritance of quantitative characters and showed how useful these methods could be. Modern views derive from the work of these experimental geneticists.

The fertile union of genetics first with plant-breeding and later with other fields of agricultural research and application occurred even at the beginning of this period. The benefits which resulted both for science and for the practice of agriculture were due to a few far-sighted persons, such as W. J. Spillman whose summary of this decade, "Application of Some of the Principles of Heredity to Plant Breeding," was issued by the U.S. Department of Agriculture, December 31, 1909. In Great Britain, Sweden, and Germany, similarly, a few persons saw the importance of the new views of heredity and some breeding institutes like that at Svalöv in Sweden became centers of genetical research at which discoveries of primary importance were made. This decade saw the first proof of Mendelian inheritance of disease resistance (Biffen, 1905), the Mendelian interpretation of the effects of inbreeding and of hybrid vigor (Shull, 1908), and, in general, a rationalization of methods of selection and of other programs of plant and animal improvement.

The General Validity of Mendel's Principles

One of the first questions which the rediscovery of Mendel raised was how general his rule was. Although Mendel relied chiefly on the results of his experiments with peas, he confirmed these with crosses of bean varieties which showed some additional complications (1866). In a letter to Naegeli, July 3, 1870, he wrote: "Of the experiments of previous years those dealing with *Matthiola annua*, *M. glabra*, Zea and Mirabilis were concluded last year. Their hybrids behave exactly like those of Pisum." Moreover, he could find no contradiction of his principle of independent assortment in the vast collection of data on hybrids published by Gaertner. Nevertheless, he made no broad claims of application beyond the actual experimental data. He always referred to "the law formulated for Pisum." That he expected it to hold for other species there can be no doubt, and his letters to Naegeli describe a long list of other species with which he was working. As I noted earlier (p. 14), many of these were in the genus Hieracium, in which Naegeli was the chief specialist, and it was with just this material that Mendel met his chief defeat. Although Mendel is said to have expressed faith in the eventual recognition of his "law" as important and general, there were no signs of this happening at the time of his death in 1884.

Two of those who rediscovered the law of segregation demonstrated it in plant species other than Pisum—eleven species in the case of de Vries; maize and peas in Correns' experiments—but neither supposed the law to be of universal application.

The first confirmation with animals by Bateson working with fowls, and Cuénot with mice, has already been mentioned (p. 66). From 1902 to 1911, Cuénot published a series of papers reporting his study in which the inheritance of coat colors of house mice was analyzed by experimental breeding methods. In addition to supplying a Mendelian interpretation of the heredity of the common coat-color variations, some of the complexities of gene interaction were explained and concepts of lethal genes and multiple allelomorphism were introduced. W. Haacke had already begun to study these same problems about 1890 and his full data, published in 1906 (*cf.* Lang, 1914, pp. 498–502), were in general agreement with Mendelian expectations. Likewise, G. von Guaita had published in 1898–99 results of experimental breeding which, after the rediscovery, could be interpreted as showing that the waltzing gait in the mouse was inherited as a simple recessive to the normal allele.

We have already seen that Bateson in 1902 had interpreted Garrod's (1901) findings on alkaptonuria as due to a recessive gene. Other human traits such as brachydactyly (Farabee,

1905) were soon shown to follow the same rule. At the end of the period, in 1910, E. von Dungern and Ludwik Hirszfeld showed that human ABO blood groups were genetically determined.

In 1902, Castle began a systematic study of heredity and variation in laboratory rodents which continued for over fifty years, and, in 1905, he introduced Drosophila as an experimental animal, both of which innovations had wide influence on the development of genetics.

Results of research on other animals came quickly and were reviewed by Bateson in 1902–1908 in his "Reports to the Evolution Committee." In 1914 appeared the first volume of Arnold Lang's ambitious work, *Experimentelle Vererbungslehre in der Zoologie seit 1900,* so extensive that in 890 pages of the first volume he had treated only the mammals. Rodents alone required 230 pages. Thereafter, the accumulation of published literature in animal genetics was so large that neither Lang nor anyone else attempted to review it. By 1910, it was in any case evident that the basic principles of segregation and recombination were of general application in higher animals including man as well as in plants.

– Chapter 9 –

CLARIFYING IDEAS:
JOHANNSEN, CASTLE, AND WOLTERECK

THE DEVELOPMENT OF GENETICS owes a great deal to two sets of ideas introduced in the period just following the rediscovery of Mendel's principles. Both deal with the interrelations of heredity and environment and both had important effects in clarifying the ideas underlying the applications of genetics to the study of evolution. One was Johannsen's fundamental distinction between genetic constitution and its expression in the characters of the individual—the distinction between genotype and phenotype. The other was the concept of the *"Reaktionsnorm"* ("norm of reaction" or "range of reaction") introduced by Woltereck in the same year.

Genotype and Phenotype

Wilhelm Ludwig Johannsen (1857–1927) had been attracted to the study of heredity and variation in the 1880s. He had begun as a pharmacist's apprentice, and had taken the pharmacy examination in 1879. He then attended lectures in botany at the University of Copenhagen and, in 1881, became an assistant in chemistry under Kjeldahl at the newly founded Carlsberg Laboratory in Copenhagen. There he began the

study of dormancy in buds and discovered methods of break-
ing dormancy by ether. (Sixty years later this was exploited
as new.) His discovery led him into plant physiology, which
he studied with Eugenius Warming, whose textbook of general botany he revised. Johannsen himself wrote a textbook of
plant physiology "with special reference to plant-cultivation."
Then, in 1892, although he had no university degree (he never
took one) he began to lecture in botany and plant physiology
at the Danish Royal Veterinary and Agricultural College. Just
when he read Galton's "Theory of Heredity" of 1876 is not
clear, but he seems to have been deeply impressed both by
Galton's so-called law of regression and chiefly by the biometrical methods introduced by Galton and developed further
in his *Natural Inheritance* of 1889.

Johannsen began the systematic study of quantitative variation in plants in the 1890's, first under the influence of Darwin, Galton, and de Vries, later of Bateson. The first results of
his new direction of work appeared in 1896 in a little book
on heredity and variation (*Om Arvelighed og Variabilitet*)
addressed to the students and published by the Student Association of the Agricultural College. His book gives a remarkable example of erudition and of logical discussion of the
problems of evolution and reveals a stimulating awareness of
the unsolved problems of heredity. Galton and Darwin excited
his interest in heredity, but it had already been aroused by
Maupertuis' observations of 1751 to which the French scientist (discoverer also of the "principle of least action") applied
probability estimates and from which he derived conclusions
concerning the origin and transmission of a human six-fingered
variant to several generations of descendants (*cf.* Glass, 1947).

Apparently, Johannsen's first purpose was to test the validity
of Galton's law of filial regression. This stated that offspring
in their quantitative characters, such as size, tend to express
the average of the race to which they belong rather than the
average of their own parents. Accordingly, very tall parents
would tend to have children less tall than themselves and very
short parents, children less short.

Johannsen set out to test this law with one of the same characters which Galton had used in deriving it. Galton had first used sweet-pea seeds. Johannsen chose varieties of edible beans and grew them in pedigree cultures. He measured length, breadth, and weight of seed in parents and descendants, and analyzed the results statistically. The results of three years of experimental observations on beans were presented at the meeting of February 6, 1903 of the Royal Danish Scientific Society and published in its proceedings (No. 3) in that year under the title "Om Arvelighed i Samfund og i rene Linier" and later in German "Über Erblichkeit in Populationen und in reinen Linien," Jena, 1903 (English translation in Peters, 1959).

What Johannsen had found, in summary, was that selection within a general population of a self-fertilizing plant, such as for seed weight or length or breadth in beans, was effective in shifting the average of the population in the direction of selection. But when selection was similarly applied to the offspring derived from self-fertilization of a single plant, it was ineffective in changing the average value of the character selected. This was interpreted by Johannsen to mean that a "population" consisted of "pure lines" differing in genetic constitution. Selection applied to a mixed population sorts out pure lines, large, small or intermediate; but within a pure line, which is homogeneous since derived by generations of self-fertilization, there is no genetic variation but only fluctuation due to chance environmental differences such as position of the bean in the pod, the pod on the plant, or the plant in the field. Selection applied to such variations is ineffective, consequently the average of these in the offspring is like that of the parental pure line. The offspring of extreme variants in a pure line show regression to the average of the parental pure line; but Galton's law is inapplicable to mixed populations.

Johannsen pointed out in clear language the implications of the pure-line theory. Lamarck's claim that environmental modifications are transmitted was excluded. Hence the environment has no directing influence, and the causes of evolutionary change have to be sought in the hereditary differences between pure lines. These differences, according to Johannsen, arise only by mutation. This conclusion would support Bateson and de Vries' claims of the importance of discontinuous variations. Selection would thus be effective only in sorting out combinations of already existing mutants; it could not produce anything new.

Another important implication of the pure-line theory was pointed out in Johannsen's paper and developed further in his book of 1909 (p. 289):

> The personal qualities of the ancestry whether the immediate parents or more distant ancestors have (in my material) no influence on the average character of the descendants. But it is the type [*i.e.*, the genetic constitution] of the pure line which determines the average character of the individual, of course in cooperation with the influence of the environment in the particular place and time.

Johannsen came back to this point in his book of 1905, *Arvelighedslaerens Elementer* (p. 82):

> The individual's worth as ancestor is essentially determined by the "type" to which he belongs and not by his purely personal condition. To this main point of the theory of heredity we come back time and again.

Out of Johannsen's recognition of this as a "main point" grew the distinction stated in his book in 1909 (pp. 162–163):

> The "type" in Quetelet's sense is a superficial phenomenon which can be deceptive. Only further study will decide whether only one or, as more often the case, several biologically different types are present [in a population]. Therefore I have designated a statistical, *i.e.* purely descriptively established type, as an "appearance type" [*Erscheinungstypus*] a phenotype. Phenotypes are measurable realities, just what can be observed as characteristic, in variation distributions of the "typical" measurement, the center around which the variants group themselves.

Through the term phenotype the necessary reservation is made, that the appearance itself permits no further conclusion to be drawn. A given phenotype may be the expression of a biological unit, but it does not need to be. The phenotypes recognized in nature by statistical investigation are in most cases expressions of such a unity.

The word phenotype, however, finds its use not merely in statistically ascertained "typical" averages but can without addition be used to designate the personal peculiarities of any individual whatever. The phenotype of an individual is thus the sum total of all of his expressed characters. The single organism, the individual plant, an animal, a man, "What he is and what he does," [wie er geht und steht] has its phenotype, *i.e.* he appears as a totality of characters which are determined by interplay between "inherited predispositions [*Anlage*] and elements of the environment."

It is significant that Johannsen defined first the concept of "phenotype," for that is the observed reality. He then sought a word to replace *Anlage,* which in German, as in the Danish *"Anlaegg,"* has many meanings. He finally settled on the last syllable of the word "pangene" which de Vries took from Darwin's pangenesis. He then combined "gene" with "type" to make "genotype" and thus achieved a counterpart to phenotype. Whereas the phenotype is an organic whole, reached by interaction between the parts of the developing organism and between these and the changing environment, the foundation from which the development begins is given by the constitution of the two gametes which unite to produce the individual (1909, pp. 165–170):

This constitution we designate by the word genotype. The word is entirely independent of any hypothesis; it is *fact*, not hypothesis that different zygotes arising by fertilization can thereby have different qualities, that, even under quite similar conditions of life, phenotypically diverse individuals can develop.

Then, pointing out that the genotype consists of elements, genes, which are, in theory, separable but in practice sometimes not separable, he goes on to give his view of what a gene is:

> The gene is thus to be used as a kind of accounting or calculating unit [*Rechnungseinheit*]. By no means have we the right to define the gene as a morphological structure in the sense of Darwin's gemmules or biophores or determinants or other speculative morphological concepts of that kind. Nor have we any right to conceive that each special gene (or a special kind of genes) corresponds to a particular phenotypic unit-character or (as morphologists like to say) a "trait" of the developed organism.

Johannsen saw more clearly than most of his contemporaries how necessary it was to restrict oneself to what we now call operative definitions, and to avoid conceptions not derived from experimental evidence. Barthelmess has cited an especially striking expression of this from Johannsen's 1909 book (p. 485):

> The conception of the gene as an organoid, a little body with independent life and similar attributes, is no longer to be considered. Assumptions which would make such a conception necessary, fail utterly. Putting a horse in the locomotive as a cause of its motion—to use Lange's classical example—is just as "scientific" an hypothesis as the organoid "explanation" of heredity.

The direction in which genetics developed as a radically experimental science, and the speed with which it moved owed much to Johannsen's vigor and critical sense. He not only foreshadowed the use of analytical statistical methods as applied to genetical problems (the first two-thirds of his 1905 *Elementer* was devoted to such methods); he also supplied another ingredient most necessary in a developing (or a developed) science: a system of simple logic applicable to its special problems.

One of the final paragraphs of his lectures on heredity (1905, p. 244) is worth quoting:

> The great progress which the introduction of statistical methods brought to the study of heredity is to be attrib-

uted unconditionally to statisticians with Galton in the lead; that Mendel independently applied statistical treatment to his hybridization results, had in fact no influence on the development of the study of heredity. Mathematicians have thus reformed the study of heredity; but they cannot and must not lead it.

This is the attitude he infused not only into his students but into many European, especially German, biologists. He expressed it succinctly in his textbook of 1909, *Elemente der exakten Erblichkeitslehre*, in the often-quoted sentence, "We must pursue genetics *with* mathematics, not *as* mathematics." His insistence that the road to the study of heredity and variation must lead through statistics and probability theory, carried the doctrines of the English school founded by Galton to the Continent, where, thanks to Johannsen among others, biologists were less handicapped by pre-Mendelian conceptions of continuous variation and blending inheritance.

Likewise, Johannsen, in effect, cleared the air of the fear that acquired characteristics might, after all, be inherited. Weismann's arguments had in the long run been less effective than Johannsen's simple experimental demonstrations, at least with those biologists who wanted to advance the study of heredity. Johannsen's conclusion that acquired modifications were not inherited was backed up a little later by W. E. Castle and John C. Phillips (1909), using an argument of quite a different kind.

There had been persistent doubts throughout the early period of Mendelism that the hereditary factors (genes) might be subject to modification by influences acting upon them through the body of the parent. If this were to happen, the "purity" of the germ cells could be "contaminated," the effects of the environment could become cumulative, and evolutionary changes could be directed by such influences. Such doubts were dispelled by a simple and decisive experiment performed by Castle and Phillips. They transplanted the ovaries from an immature black guinea pig to an albino guinea pig whose ovaries had been operatively removed. The albino with the ovaries from the black was then mated to an albino male.

Castle had already shown that albinism was inherited as a simple recessive to black. If the germ cells of the black animal were to be affected by the albino character of the foster mother into which the black ovary had been implanted, this should be revealed in the progeny arising from these germ cells. However, the albino foster mother when mated to an albino male bore three litters consisting entirely of fully black offspring. No influence of the albino foster mother, in whose body the eggs had been produced, could be seen. In Castle's words (1951, p. 69): "This showed that the black ovary had survived unmodified and functional in an albino body, and that Weismann was right in maintaining that germ cells are distinct from body cells." This demonstration convinced many doubting biologists and made the rather radical claim of the Mendelians (that genes were subject to modification only by mutation) an acceptable assumption on which further experiments could be based.

The Norm of Reaction

Implicit in Johannsen's demonstration of two kinds of variability—genotypic due to mutation and phenotypic due to response of the genotype to the environment—was an idea which was shortly to be made explicit. Richard Woltereck, professor of zoology at Leipzig, published between 1908 and 1911 a series of papers reporting quantitative studies of variation and inheritance in small freshwater plankton crustaceans belonging to the genus Daphnia. Since an important method of reproduction in these animals is by parthenogenesis, Woltereck could isolate pure lines and confirm in animals the main conclusions of Johannsen which were based on self-fertilizing plants. At the same time, Herbert S. Jennings was observing variation within and between pure lines in Paramecium reproducing by fission and later in the shelled protozoan Difflugia. Woltereck's interest centered in the problem which had been brought into sharp focus by the discoveries of Mendel (segregation) and de Vries (mutation) as applied by Johannsen to the causes of species formation or alteration.

The new view then emerging—that continuously variable modifications could not furnish the materials for evolutionary change—seemed to Woltereck to be irreconcilable with Lamarckian views then current in botany, as well as with those generally held by zoologists whether Lamarckians or selectionists. He stated these opposing views as follows (1909, p. 113): "old: continuous species change determined by effects of environment; new: discontinuous changes without determination by the environment."

The triumph of the new forecast by the work of de Vries and Johannsen seemed likely, as Woltereck put it, to deal "hard blows against just those conceptions of the nature of the organic world in which we have all grown up."

Without special reference to the species problem as such, Woltereck proposed an attempt to analyze the mode of interaction of genotypic variations with environmental ones. He thus studied quantitatively varying and readily measurable characters, chiefly head form, within and between pure lines of daphnids. These were reared for a succession of generations in controlled environments with measured differences in temperature and nutritive level. Woltereck obtained for each population under each condition a frequency distribution of the character under study. He called this a "phenotype curve" and introduced the new idea as follows: "The numerical relations as a whole represented in all of these curves can be designated as the specific and relative *norm of reaction* of the quantitative characters being analyzed . . ."[19] The position to be defended in the discussion is then stated (1909, p. 136): "The genotype [the genotypic foundation] of a quantitative character is the inherited norm of reaction." This statement rests on two propositions: (1) that which distinguishes one biotype or pure line from another is passed on in the gametes from one generation to the next; (2) a new line originates by way of an hereditary change in some specific trait. The

---

[19] "Die in all diesen Kurven dargestellte Gesamtheit der Relationszahlen können wir als die spezifisch-relative *Reaktionsnorm* des analysierten Quantitativmerkmals bezeichnen. . . ."

Daphnia studies unquestionably support the following (1909, p. 136) :

1. The total norm of reaction with all its countless specific relations is inherited as a head-form predisposition [Helm anlage].

2. The biotypes of Daphnia arise and have arisen through hereditary changes in the norm of reaction of its head height and its other traits.

Thus arose a clear view of the relation between heredity and environment which, had it been more generally appreciated, would have led to more rapid progress in the development of evolutionary biology and especially in understanding the bases of human variation. As one of Woltereck's listeners said after the lecture, "Nihil est in selectione, quod non fuerit in variatione" ("There is nothing in selection that did not first appear through mutation"). This was, perhaps, to remind his colleagues of the German Zoological Society, gathered for the 100th anniversary commemoration of Darwin's birth (of which Woltereck's lecture was a part), that natural selection by itself can create nothing new.

— Chapter 10 —

RECONCILING APPARENT EXCEPTIONS
TO MENDEL'S RULES

THE EFFECT of the rediscovery of 1900 was, as we have seen, to set off a kind of chain reaction which led to the application of Mendel's methods of pedigree breeding to the analysis of heredity and variation in many species of plants and animals. As soon as results began to be reported, they revealed apparent departures from those results from which Mendel had derived his principles. Most of these departures proved to be due to variations in the manner in which genes express themselves in development and thus did not bear on Mendel's principles.

Absence of complete dominance has already been referred to. Correns' clear vision should have inhibited from the start the dogma of dominance which clings to Mendelism even to this day. Correns seems to have expected expression of both alleles in the hybrid and consequently looked for it even when it did not strike the eye. He usually found it. His doubts that any single trait could be fully dominant led him to compare large masses of F_1 flowers with colored and colorless parents (in *Hyoscyamus niger* × *Hyoscyamus pallidus*) and to reveal differences between hybrid and colored parents. Correns also foreshadowed later work by introducing quantitative methods of estimation of anthocyanin and other plant pigments and proved in this way the influence of the "recessive" allele on synthetic processes.

Even in 1900, Correns was aware that the same or similar traits could be due to different hereditary factors. This again provided a liberalizing influence opposing the rather mechanical view of "unit factors" with exclusive control over particular characters which continually threatened the new views of heredity.

Other cases worked out in the early years of Mendelism revealed that the expression of one mutant gene could be suppressed by the action of a different and independent one. Bateson analyzed cases in which genes determining different colors or patterns failed to affect the phenotype when a dominant gene inhibiting all color is present. This was called "epistasis" by Bateson. The gene expression which was suppressed was said to be "hypostatic." The terms proved to be useful and survived the death of the presence-and-absence hypothesis on which they had been based. They turned out to be examples of a more general phenomenon which became evident during this decade, 1900–1910, namely that the phenotype depends on the interaction of many genes with each other and with the environment and that each gene has many effects in addition to the most striking one by which it was first recognized. Failure to appreciate this fact had led to the retention of the notion of unit characters which plagued genetics for two decades and doubtless delayed the clarification of some of its basic concepts. The idea probably derives from de Vries' "unités caractères" of 1900 and this, in turn, from his *Intracellular Pangenesis* of 1889. The essential assumption was that "the specific characters of organisms are composed of separate units." When these had been demonstrated in hybridization experiments, they became the elements of character representation without reference to any material carriers. This seems to have been the sense in which Bateson (1902) took over the term. It became the equivalent of Mendel's "differentiating characters" and later of "unit factors" (the material cause of the unit character) and of "gene" as substituted for "unit factor" by Johannsen in 1909. Even after Johannsen had clarified the distinction between genotype and phenotype, the vagueness of "unit character" persisted. Erwin Baur pointed out in 1907 that what distinguished de Vries' view from Johannsen's was just the different meaning attached to "character" and "Anlage."

Spillman tried to clarify the situation in a paper published in 1910, "Mendelian Phenomena without de Vriesian Theory." His point was to transfer the emphasis from "characters," which were not transmissible as such, to the functional elements, such as enzymes, which determine characters.

As Johannsen later pointed out (1922, p. 191), the coup de grace was given to the unit-character idea by Nilsson-Ehle's demonstration in 1909 that some characters, for example, the red color of wheat grains as opposed to white, may depend for their full expression not on one but on several pairs of independently segregating alleles. In this case, a red-grained race crossed with one with colorless grains produced, in F_2, $63/64$ red and $1/64$ colorless-grained plants. Nilsson-Ehle's interpretation was that three different gene pairs R_1r_1, R_2r_2, and R_3r_3 were segregating, each pair acting in the same way on the same quality, seed-coat color. In this case, the effects of the different pairs were additive, $R_1R_1R_2R_2r_3r_3$ being redder than $R_1R_1r_2r_2r_3r_3$. This and similar cases similarly analyzed became the basis for the theory by which many cases of inheritance of quantitative characters (*i.e.*, measurable characters) were interpreted in Mendelian terms. In Europe this was called "polymeric inheritance" (*cf.* Lang, 1914). In the United States, where it was developed by E. M. East and H. K. Hayes (1911) and by George H. Shull, it was known as the "multiple-factor theory" and widely applied to interpret and guide observations on the inheritance of size, yield, height, and similar measurable traits. Kenneth Mather later referred to genes with such effects as "polygenes," but the essential meaning of all these terms is that many genes with similar quantitative effects interact with each other and with the environment in determining the course of development of the organism. The state of the characters is the outcome of this interaction and is not bound to the properties of particular genes.

In 1905, Cuénot reported the first case in which a single unit (gene) could be detected in more than the two alternative forms postulated by Mendel. Breeding experiments with mice suggested that a mutant allele for yellow coat color was allelic with both agouti and non-agouti black. R. H. Lock (1906) first clearly expressed the idea of three instead of two alternatives. Later this first series of multiple alleles was extended to include several additional members. Cuénot correctly inferred that some of the color varieties of mice differ not in the number of Mendelian units but in the kind; that is, such variations are not polymeric but due to different conditions of the same element. Series of multiple alleles were soon found by other investigators in many species of animals and plants. However, the case that provided the classical example of multiple allelomorphism was that of rabbits (Castle, 1905). The factor responsible for color in Himalayan albino rabbits (white with black points) was not the same as the factor for color in full colored (wild-type) rabbits and both factors were allelic to that in complete albinos. However, it remained for A. H. Sturtevant (1913) to give the first clear analytical statement of this and of similar cases of multiple allelism as due to alternative modifications of the same gene. The proof of this concept effectively disposed of Bateson's presence-and-absence hypothesis, since that hypothesis permitted only two alternative forms for any gene.

Cuénot's results (1905) with yellow mice also pointed to the first signs of the operation of lethal factors in animals. Later, Castle and C. C. Little (1910) proved that the yellow allele had two expressions: a dominant one on coat color, and a recessive one on viability, since yellow homozygotes died early in the embryonic state. The lethal-factor hypothesis, however, was first employed by Erwin Baur (1907) who found a yellow-leaved ("Aurea") form of snapdragon which was viable only in the heterozygous form.

It was Correns again who, as early as 1900, discovered the first "Mendelizing" disease. This was Sordago, a localized infection of the palisade cells in the four o'clock plant, *Mirabilis jalapa*. Correns' biographer, Emmy Stein (1950), tells us that this plant, which revealed so much to Correns concerning the extent of the segregation principle, was referred to by Correns as "Miserabilis" because of the difficulties caused in breeding experiments by having a single seed per fruit and requiring much room.

Correns, in 1902, was the first to state clearly the hypothesis of the physical mechanism of segregation of alleles which underlies the chromosome theory of heredity:

> Each tetrad contains bodies of both kinds, those with *A* as well as those with *a*, accordingly segregation must be carried out by nuclear division, in fact by the first division of the pollen mother cells.

This statement was obviously unknown to Sutton when, in October, 1902, he sent to the *Biological Bulletin* a similar statement usually regarded as the first on the physical basis of segregation. But before dealing in detail with the development of the chromosome theory, there are still other expansions of views about segregation to be considered.

One of the cornerstones of Mendel's theory was the assumption of *random union* of gametes uninfluenced by the genes they contained. From *Aa* × *Aa* the four combinations *AA*, *Aa*, *aA*, and *aa* should be equally frequent. Correns, in 1902, found an exception to this rule in which the frequency of union of an egg with a recessive allele *s* (sugary reserve carbohydrate in maize seed, as in sweet corn) was fertilized less frequently by a sperm with the same allele than by one with the dominant allele *S* (starchy reserve). Such cases of selective fertilization are rare in animals, but in plants many cases have been worked out in which the growth of the pollen tube which conducts the generative nucleus to the egg is affected by certain genes that it carries.

In the same investigation with maize, Correns found a case of coupling, or inheritance in association, of self-sterility and blue aleurone. This and his 1902 theory of crossing over will be discussed in the following chapter.

— Chapter 11 —

CHROMOSOMES AND HEREDITY*

DURING THE 1900-1909 DECADE, sex was shown to be a Mendelian character and was brought into relation with the visible mechanism of the sex chromosomes. It was from this beginning that the first strict proof was obtained, through non-disjunction of the sex chromosomes in Drosophila, that genes were in fact parts of chromosomes.

The idea seems to have occurred to several people that in mammals the one-to-one ratio between the sexes at birth resembles that obtained from mating a heterozygote with a homozygous recessive. Mendel himself had mentioned it in a letter to Naegeli, Sept. 27, 1870. Strasburger (1900), Bateson (1902, p. 138), and Castle (1903) all suggested that sex was determined by the nature of the uniting gametes, but it remained for Correns, who expressed the same idea in 1903, to provide the first experimental proof of it (1907). He found that in dioecious species of the plant Bryonia, the pollen is of two sorts, half male-determining, half female-determining, while all eggs, in respect to sex, are alike. The male was thus heterozygous, or, according to the terminology introduced by

* References to literature cited in this chapter are to be found in Wilson (1928).

Wilson in 1910, heterogametic, while the female was homogametic (*cf*. Wilson, 1928, p. 749).

The first sex-linked character was found in the currant moth *Abraxas grossulariata* in 1906 by Doncaster and Raynor. The results were interpreted by Bateson and Punnett (1908) and Doncaster (1908) as showing that the female was heterozygous (heterogametic) and the male, homogametic. Soon it was shown that the female was digametic in birds (Goodale, 1909) while in mammals and dipterous insects the male proved to be the digametic sex. By this time, 1910, there was no doubt that the differences between the sexes in higher animals and dioecious plants was a Mendelizing character. Correns' voice, however, was again prophetic, for he insisted in his first paper that each sex must have the potentialities of both so that what the character of the uniting gametes decided was which set of sexual characters should become functional during development. This implied what was proven later: that the development of one sex was suppressed by the other. This was shown also by A. F. Blakeslee's analysis (1904) of mating-type determination in the mold Mucor, in which conjugation occurs only between mycelia of opposite mating types, designated plus (+) and minus (−). The resulting sporangia produce either + or − spores, never both, so that here, too, one sort is suppressed. Similar indications that sex was a genetic property, subject to rules like Mendel's segregation principle, were obtained from both lower and higher plants and from diverse animals.

The connection of sex-determination with a chromosomal mechanism goes back to Henking's observation (1891) of an unpaired "chromatin-element" in spermatogenesis of a bug so that sperms of two different types were produced, half with and half without this element. However, Henking did not realize that this element was a chromosome (he called it "x") nor did he connect it with sex-determination. This remained for C. E. McClung to do in 1902, when he discovered an "accessory" unpaired chromosome in a grasshopper which he assumed to be sex-determining although he at first supposed (wrongly) that sperms with the accessory were male-determining. This set off an intensive study of chromosomes in relation to sex.

A decisive discovery was made in 1905 by Nettie M. Stevens and E. B. Wilson. Wilson showed that in several genera of hemipteran insects the female regularly has an even number of chromosomes and has one more than is found in the male. In the male, therefore, one chromosome is unpaired so that it goes to only half the sperms, and these give rise to females. The other half of the sperms have one less chromosome and these give rise to males. The chromosome which was unpaired in the male was at first called a "heterochromosome" or "allosome" in contradistinction to the other pairs, which were called "autosomes" (A). Later the unpaired chromosome was called the "X chromosome," and its unequal partner found in males of certain species the "Y chromosome" (cf. Wilson, 1909).

In the bug *Anasa tristis*, for example, it was found that the females have 22 chromosomes, the males 21, as follows (Wilson, 1905):

♀ ♂
20A + XX 20A + X

Eggs	Sperms		Fertilizations
10A + X	10A + X	10A + X with 10A + X = 20A + XX =	♀
10A + X	10A + O	10A + X " 10A = 20A + XO =	♂

In the beetle Tenebrio, Nettie M. Stevens (1905) found the following:

♀ ♂
18A + XX 18A + XY

Eggs	Sperms		Fertilizations
9A + X	9A + X	9A + X with 9A + X = 18A + XX =	♀
9A + X	9A + Y	9A + X " 9A + Y = 18A + XY =	♂

The discovery and correct interpretation of sex chromosomes in relation to sex determination was thus made inde-

pendently by two observers at nearly the same time. In his
biographical memoir of E. B. Wilson, Morgan (1940) stated
the following:

> The question is sometimes asked as to the priority of
> Stevens' and Wilson's papers. Stevens' paper was handed
> in on May 15, 1905 and printed in September of that
> year. In Wilson's paper "Studies on Chromosomes: I"
> (dated May 5, 1905; published August, 1905) he says in
> a footnote: "The discovery, referred to in a preceding
> footnote, that the spermatogonial number of Anasa is
> 21 instead of 22, again goes far to set aside the diffi-
> culties here urged. Since this paper was sent to press I
> have also learned that Dr. N. M. Stevens (by whose kind
> permission I am able to refer to her results) has inde-
> pendently discovered in a beetle, Tenebrio, a pair of un-
> equal chromosomes that are somewhat similar to the
> idiochromosomes in Hemiptera and undergo a corre-
> sponding distribution to the spermatozoa. She was able
> to determine, further, the significant fact that the small
> chromosome is present in the somatic cells of the male
> only, while in those of the female it is represented by a
> larger chromosome. These very interesting discoveries,
> now in the course of publication, afford, I think, a strong
> support to the suggestion made above; and when con-
> sidered in connection with the comparison I have drawn
> between the idiochromosomes and the accessory show
> that McClung's hypothesis may, in the end, prove to be
> well founded."

In Wilson's "Studies on Chromosomes: II" (dated October
4, 1905; published November, 1905), he says:

> During the summer, and since the foregoing paper was
> entirely completed in its present form, I have obtained
> new material which shows decisively that the theoretic
> expectation in regard to the relations of the nuclei in the
> two sexes, stated at p. 539, is realized in the facts. In
> Anasa, precisely in accordance with the expectation, the
> oogonial divisions show with great clearness one more
> chromosome than the spermatogonial, namely, twenty-two
> instead of twenty-one; and the same number occurs in
> the divisions of the ovarian follicle-cells. Again in accord-

ance with the expectation, the oogonial groups show four large chromosomes instead of the three that are present in the spermatogonial groups. In other respects the male and female groups are closely similar. In like manner, the oogonial divisions in Alydus and Protenor show fourteen chromosomes, the spermatogonial but thirteen; and in Protenor the spermatogonial chromosome-groups have but one large chromosome (unquestionably the heterotropic) while the oogonial groups have two such chromosomes of equal size.

In other cases of insects in which males and females had the same number of chromosomes, sex determination was effected by the XY segregation as first shown by Stevens. Later in Lepidoptera and birds, XY ♀ and XX ♂ were discovered, but only after the distribution of sex-linked traits suggested that females were heterogametic.

Morgan, in 1910, concluded that although there would be difficulties in reconciling all the evidence, what had been proved was that there was in the germ cells a mechanism connected with sex and that the external conditions, heretofore often invoked as sex-determining, were in fact not decisive.

Chromosomes and Heredity

The above heading was the title of the review (1910) in which Morgan summed up the progress of the decade in respect to what later came to be referred to as "The Chromosome Theory of Heredity"; that is, the view developed largely by Morgan, Wilson, and their school that chromosomes consist of linear arrangements of genes in an order that can be mapped by experimental breeding methods.

Two other sets of ideas had previously been called the chromosome theory. One was the speculative system arrived at deductively by Weismann, which he and Roux developed in the 1880s. The other was developed by Boveri, who was often cited as the founder of the chromosome theory.

Weismann's earlier theory differed from Boveri's later one

not only in the manner it was derived but also in the nature of the elements assumed to lie in the chromosomes and especially in the relation of the elements to development and differentiation. It was the changes in the then newly discovered "chromatic bodies" (*i.e.*, chromosomes) during the maturation divisions of the germ cells that set off Weismann's train of thought, which as Morgan later said "acted like yeast in the minds of less imaginative workers." The chromosomes were viewed as collections of "determinants" for organs and body parts which were passed into the germ cells from the "germ line" and, in the developing embryo, were distributed and parceled out by qualitatively different mitoses, so that different parts got different assortments of chromosomes and determinants.

Boveri's chromosome theory, on the other hand, was based on the fact that he had been at such pains to prove—namely, that each chromosome pair differs from every other one in the qualities it carries. "Chromosome individuality" was Boveri's expression for this view, but as Wilson has pointed out (1928, p. 828) the essential evidence for persistent individuality is that "every chromosome which issues from a nucleus has some kind of direct genetic connection with a corresponding chromosome that has previously entered that nucleus." Genetic continuity of each chromosome throughout the life cycle is thus the primary cytological principle which Boveri developed.

The Sutton-Boveri Hypothesis

In December, 1902, a young graduate student at Columbia University, Walter S. Sutton (1877–1916) had a paper, "On the Morphology of the Chromosome Group in *Brachystola magna*," published in *The Biological Bulletin*. Its essential feature was evidence from cytological observation that the chromosomes that pair in synapsis consist, in each pair, of one maternal and one paternal member. T. H. Montgomery had already reached this conclusion by studying chromosomes of other insects, but Sutton had confirmed and extended it and added to it a new and original idea. The last sentences of his 1902 paper were as follows:

> I may finally call attention to the probability that the association of paternal and maternal chromosomes in pairs and their subsequent separation during the reducing division as indicated above may constitute the physical basis of the Mendelian law of heredity. To this subject I hope soon to return in another place.

He returned to it in January, 1903, when he sent a second paper to *The Biological Bulletin,* "The Chromosomes in Heredity," which appeared in April, 1903. In this, he outlined in remarkably clear form what he referred to as the chromosome theory of Mendelian heredity which resembled in most respects the interpretation reached experimentally by Morgan, Sturtevant, Muller, and Bridges and reported in the *Mechanism of Mendelian Heredity* (1915). It was based entirely on cytological observation and interpretation, by which Sutton explained the then known facts of Mendelian heredity. This was the first consistent application of observations on chromosomes to the results of experimental breeding. "Cytogenetics," a term invented much later, may be said to have begun with Sutton's paper.

The bases of Sutton's theory were:

1. There exists, in diploid organisms, a chromosome group consisting of pairs of homologous chromosomes, one from the father and one from the mother, in each pair.
2. The homologues pair at synapsis.
3. One homologue of each pair enters each gamete.
4. The distribution of members of each pair at meiosis is independent of that of every other pair.
5. Each chromosome pair is different (in Sutton's case,

     morphologically different) from every other pair and retains this individuality throughout all mitoses and meioses.

6. Genes are parts of chromosomes. (For genes, Sutton used the newly invented term "allelomorphs" or unit characters.)

What Sutton proposed, in effect, was a theory based on the parallelism between the distribution of pairs of chromosomes and the distribution of segregating pairs of alleles at meiosis. He supposed that some chromosomes at least are "related to" a number of different allelomorphs which may be dominant or recessive independently (p. 240), but he did not allow for recombination (crossing over) between allelomorphs in the same chromosome and so supposed that "all allelomorphs represented by one chromosome must be inherited together."

How ripe the cytological facts were for application to Mendelian heredity is shown by the list of persons who expressed ideas similar, in some degree, to Sutton's: Strasburger in 1900 and 1901, Correns in 1902, and W. A. Cannon in 1902 (*cf.* Sutton, 1903; Wilson, 1928, pp. 923–928). Boveri had noted in 1902 the connection between chromosome behavior and the plant hybridization experiments (*vide* Sutton, 1903), but his full discussion was delayed until 1904. Cannon's paper was titled "A Cytological Basis for the Mendelian Laws," but it contained, as Sutton pointed out, the fundamental error of assuming that the maternal and paternal homologues were segregated as intact groups into different germ cells. It is the three men: Correns (1902a), independently of Sutton, de Vries (1903) who based his theory mainly on Sutton's but improved on it, and Boveri (1904) who can be bracketed with Sutton as responsible for launching the cytological interpretation of Mendelism.

Correns (1902a) assumed, independently of Sutton, that

the alleles (*anlagen*) lie in serial order, one set in each of the homologues of a hybrid as:

$$\frac{\text{a b c d e}}{\text{A B C D E.}}$$

Recombination may then occur by rotation on the long axis of one or more pairs to produce orders like:

$$\frac{\text{a B c D e}}{\text{A b C d E}} \quad \text{or} \quad \frac{\text{A B c d E}}{\text{a b C D e}}$$

This looks superficially like the kind of recombination later imputed to crossing over with formation of chiasmata between homologues; but the latter process has the essential difference of involving the exchange of blocks of alleles, a point which Morgan always stressed. Moreover, the order of *anlage* (or loci) has to be constant if the data on recombination are to be accounted for, and Correns' theory did not envisage this.

However, de Vries, in 1903, clearly assumed that the exchange of *anlage* takes place between maternal and paternal homologues at synapsis. This assumes, as Wilson pointed out, that the individual units must lie in the same serial order, with alleles opposite each other. De Vries considered his theoretical paper (presented at the 151st meeting of the Netherlands Scientific Association and titled "Befruchtung und Bastardierung"—"Fertilization and Hybridization") of such importance that he arranged for an English translation of it to be included in the American edition of *Intracellular Pangenesis* (1910). De Vries' conception was largely based on Sutton's paper, but while it went beyond it in suggesting a theory of crossing over, the assumed exchanges, as in Correns' theory, involved only single alleles and not the segments of chromosomes which formed so essential a part of Morgan's theory.

Boveri had already reached views similar to those of Sutton, although they were not published until 1904 and then without the detailed application to Mendelism which Sutton had proposed. But there is no doubt that it was Boveri who proved experimentally the qualitative, that is the functional, individuality of the chromosomes as essential elements in the control of development. This was an idea more essential to the eventual theory than the morphological distinctions which Sutton relied upon. Said Boveri (1903, p. 30) :

> The more our insight grows, the more we perceive that "morphological" in these matters is only the sub-structure of what we want eventually to know: what in fact these chromatin elements, which bear such remarkable destinies, possess in respect of physiological significance.

John A. Moore (1963) has given a clear and readable account of the work of both Boveri and Sutton, while a more detailed discussion based on Boveri's correspondence as well as his published work is in the excellent life of Boveri by his student Fritz Baltzer (1962). Wilson (1928), whose book was dedicated to Boveri, called the definitive ideas connecting the chromosome mechanism with Mendelian heredity "the Sutton-Boveri hypothesis." It was Wilson, the sponsor of Sutton as a graduate student, who testified to the originality of Sutton's basic ideas. Victor McKusick (1960) has given us a documented account of Sutton's short life and his relation with Wilson and with his first teacher, C. E. McClung, at the University of Kansas. He eventually took his degree in medicine and became a surgeon. He died in 1916 at 39 years of age without having returned to the field of his two classical publications.

Linkage

One of the discoveries leading to the chromosome theory was made entirely without reference to it. It was presented in the report (1905) by Bateson, Saunders, and Punnett on linkage, seen in partial coupling in sweet peas of the genes for long pollen shape and colored flowers. This discovery was clearly established in 1906 and it was assumed, in explanation, that the gametes with the coupled characters received from the parent, *e.g.* *AB* and *ab*, had undergone additional equational divisions after reduction (reduplication hypothesis). Although this was elaborated further and made to cover the contrary cases of repulsion, *i.e.* additional mitotic division of *Ab* and *aB* gametes when the characters had come from the parents in these combinations, the hypothesis was badly strained and unable to account for gametic series in ratios other than $2^n : 1 : 1 : 2^n$, n being the number of additional post-segregation equational divisions. This hypothesis became unnecessary after Morgan's proof that linked genes are held together by being parts of the same chromosome.

The Chiasmatype Theory of Crossing Over

The two main events of the last half of this decade in respect to the chromosome theory were (1) the beginning of T. H. Morgan's work in genetics and (2) the publication, in 1909, of "La Théorie de la Chiasmatypie, Nouvelle Interpretation des Cinèses de Maturation" by the Belgian cytologist F. A. Janssens. This paper was read at the May 5, 1908, meeting of the Academie Royale de Belgique, but publication was postponed by Janssens for over a year. Janssens had observed in the chromosome preparations at meiosis of certain amphibians the occurrence of cross figures at synapsis, to which he gave the name "chiasmata." He interpreted each of these as due to a fusion at one point between two of the four strands of the tetrad, followed by breakage and reunion leading to an exchange of equal and corresponding regions of two of the four chromatids. The maternal chromosome of the synapsing pair was assumed to have replicated at this time and the paternal chromosome also, so that a chiasma represented a point of change of partner between a maternal and a paternal chromosome. This would account for recombination of genes in the same chromosome. It was this feature that Morgan seized upon in constructing his theory of linkage with crossing over between blocks or groups of genes, which became the cornerstone of his chromosome theory. It was the opinion of E. B. Wilson (1928, p. 960) and of other cytologists, *e.g.* C. D. Darlington (1937, p. 253), that the cytological evidence provided by Janssens was not sufficient to prove that exchange (crossing over) did in fact occur. But when Morgan, in 1911, provided incontrovertible evidence of recombination between two sex-linked genes in Drosophila, the resulting chiasmatype theory of crossing over united Janssens' cytological with Morgan's genetical evidence and ideas. This became the basis for the elaborate and fruitful development of the theory of the linear order of gene loci in chromosomes, as summarized by Morgan in 1926. The importance of Janssens' work lay not in what it proved cytologically but in what it suggested for testing by genetical methods. Wilson stated it in this way in 1920 (Wilson and Morgan, 1920, p. 210):

> I am not able to escape the conviction that somewhere in the course of meiosis some such process must take place as is postulated by Janssens and by Morgan and his co-workers, though I must admit that this opinion rests less on cytological evidence than on genetic.

The development of the theory of the "Mechanism of Mendelian Heredity," based first on work with Drosophila by Morgan, Sturtevant, Muller, and Bridges belongs, in fact, to the next decade. Although Morgan's Drosophila breeding began in 1909, the first discoveries were reported in the spring of 1910.

– Chapter 12 –

THE BEGINNINGS OF
POPULATION GENETICS: 1869–1910

INTEREST IN THE CAUSES of change in natural populations and in the directed evolution of races of domesticated animals and plants was in evidence long before it reached organized expression in the theories and methods of population genetics. Darwin had focused attention on both of these subjects, first in his theory of natural selection as the guiding force of evolution, and then in the evidence by which he supported it drawn from the variation of animals and plants under domestication. During the decade 1860–69, two lines can be discerned which were later to fuse in modern population genetics.

One of these lines was formed in the work of Francis Galton and led through the work of Karl Pearson and G. U. Yule to the development of biometry. Nilsson-Ehle, East, and Shull supplied essential proof for the multiple-factor theory and this led to quantitative genetics.

The other line of development began with Mendel (1866) and led to the analysis of the effects of inbreeding and cross-breeding and of Mendelian equilibrium in random-mating populations which form the rational bases for population genetics and for genetic interpretations of processes of evolution.

Galton, in his *Hereditary Genius* of 1869, conceived of heredity as "concerned with Fraternities and large Populations rather than with individuals." He therefore believed that the quantitative statistical study of metrical characters in populations could lead to a general theory of inheritance. This attitude continued to distinguish the ideas and methods of the biometric school descended through Galton, from those derived from Mendel and based on the effects in inheritance of differences between individuals. Reconciliation between the two views came only after the bitter battles between the English "biometricians" and "Mendelians" in the early 1900s revealed to other biologists that the two views were in fact complementary and aspects of the same basic mechanism.

The biometrical view was first thoroughly developed in an important book by Galton, *Natural Inheritance* (1889). It was based on his observations that (1) human stature is influenced by heredity, there being a correlation of about .33 between parent and child (later shown to be closer to .5); (2) "the characteristics of any population that is in harmony with its environment, may remain statistically identical during successive generations" (1889, p. 192). He reconciled these two facts by his law of universal regression: that is, that individual peculiarities (defined as the degree to which a personal attribute differs from that of the population) are shared by relatives, but *on the average* to a lesser degree. Children resemble parents but regress toward the population average and so also with more distant relatives between which the degree of resemblance as measured by the correlation coefficient declines toward zero. In direct ancestry he assumed about $\frac{1}{4}$ of the heritage of each child to have come from each parent, $\frac{1}{16}$ from each grandparent, etc. In respect to stature, he supposed that the heritage was composed of small units "too minute for its elements to be distinguished" and thus giving the appearance of blending when the parents were of different heights. The treatment throughout is statistically controlled by methods invented for the purpose; in fact, this may be said to be the beginning of modern biometry.

It had already been pointed out in 1867 that if blending inheritance were the rule, then the incorporation of a new and better variation into a population could hardly take place as Darwin had supposed it should in his theory of natural selection. This criticism was made in an article on the *Origin of Species* by Henry Charles Fleeming Jenkin (1833–1885) published in the *North British Review* in June, 1867. Jenkin's main point (p. 294) was that if a new and more fit variation was to arise as a sport (equivalent to what was later to be called a mutation) then "the sport will be swamped by numbers and after a few generations its peculiarity will be obliterated." Francis Darwin, in his *Life and Letters of Charles Darwin* (1887, p. 288), wrote:

> It is not a little remarkable that the criticisms, which my father, as I believe, felt to be the most valuable ever made by his views, should have come, not from a professed naturalist but from a Professor of Engineering.

Jenkins' "argument from swamping" proved to be fatal, not for the theory of natural selection (although Darwin was unable to meet Jenkin's objection on this score) but for the theory of blending inheritance. The resolution of the difficulty had, of course, already been foreshadowed by Mendel's discovery of the principle of segregation in 1865, two years before Jenkin's publication. If elements such as genes assume different forms by mutation and retain their integrity in all combinations, then new variants so arising cannot be swamped out. On the contrary, they will be retained in the population which, as S. S. Chetverikov (1926) later expressed it, "absorbs them like a sponge." This consequence of Mendel's principle, also emphasized by R. A. Fisher in 1930, underlies the application of genetics to the interpretation of evolutionary changes in populations.

W. E. Castle (1903) worked out the expected distribution of a recessive gene in a population mating at random and found no change in frequency over several generations. This was a first adumbration of the equilibrium principle, as A. H. Sturtevant and Curt Stern pointed out at a meeting of the American Philosophical Society in April, 1965.

An important landmark in the development of ideas in both biometrical and Mendelian genetics was Karl Pearson's paper of 1904, "On a Generalized Theory of Alternative Inheritance with Special Reference to Mendel's Laws." In this paper, he both rejected Mendelism and at the same time correctly generalized the principle of segregation showing that the F_2 ratio ¼ AA : ½ Aa : ¼ aa should maintain itself indefinitely in a large, random-breeding population. This was an explicit statement of the equilibrium principle for single loci, and its application to multiple loci could have been inferred from this. Pearson's rejection of Mendelism was based on a discrepancy between the value of .33, which he found for the genetic correlation between parent and offspring, assuming segregation and complete dominance, and the observed values found for characters like stature, which were much higher, of the order of .5. Therefore, he said, the Mendelian interpretation is excluded. G. U. Yule (1906) soon pointed out that with intermediate dominance, as possible with Mendelian inheritance, the genetic correlation would be .5 as observed. But, in 1908, Pearson could still say, "Mendelism has not been demonstrated for any one character" (cf. Dunn, 1962).

It was this failure of the biometric school descended from Galton to appreciate the general validity and the implications of Mendelian heredity which excluded them from the advancing front of genetics for this period and prevented their exploitation of the field which they had themselves opened up. The fact was that a mechanism of particulate heredity based on genes could not be deduced from correlations between relatives; whereas the correlations between relatives found by biometricians were necessary consequences of Mendelian heredity. However, as Sewall Wright (1960) has pointed out, the bridge between biometrical population genetics and Mendelian experimental genetics had in fact been constructed, although inadvertently, by Pearson in 1904. The extension of the bridge by Hardy and Weinberg in 1908 and Weinberg's synthesis of 1910 will be discussed after the part played by Mendel and his successors has been outlined.

Biometry

In spite of the temporary opposition of biometricians to Mendelism, which was largely confined to the English school under Pearson, they nevertheless opened a fertile field in the development of methods for dealing with continuous, quantitative variation. The rationale, to be sure, was based on a wrong assumption—that continuity of variation of characters was an expression of continuous or blending genetic variation. But, nevertheless, such properties of the normal or Gaussian "curve of errors" as were useful in analyzing data from populations—standard error, coefficient of variation, goodness of fit by the chi-square criterion, coefficient of correlation—these were developed and applied with great vigor.

Karl Pearson, who began as an economist but entered biometry as a protégé of Galton's, was the leader in this work. He, together with Galton and Weldon, founded the journal *Biometrika* which began publication in 1901 and soon became the chief repository of biometrical data and methods of analyzing quantitative variation.

A great impetus was given by Johannsen's *Elemente der exakten Erblichkeitslehre* when it appeared in a German edition in 1909. It gave clear statements of statistical methods with examples of their use in genetical problems; but, of even greater importance, Johannsen succeeded in removing the basic cause of contention between biometricians and Mendelians. He showed that breeding tests, and only these, were valid bases for distinguishing between inherited variations due to gene differences and continuous variations due to nongenetic causes, *i.e.* the pure-line methods and the concepts of phenotype and genotype which arose out of them. With the proof by Nilsson-Ehle, and independently by East (1910), that continuous hereditary variation could be explained by discontinuous genetical variations (*i.e.*, genes altered by mutation), the last of the grounds for division between biometry and Mendelian genetics was removed and thereafter both advanced together. The state of their relations in 1910 was well described by Raymond Pearl (1911 and 1915).

Mendelism and Population Genetics

Two sets of ideas which underlie the development of population genetics were derived directly from Mendel's paper. One of them, which led to the theories of inbreeding, was generalized by Mendel himself. In the section of his 1866 paper headed "The Subsequent Generations Bred from the Hybrids" he pointed out that one of the automatic consequences of segregation in self-fertilized organisms, such as peas, was the orderly and regular decrease of the proportion of heterozygotes in successive generations derived from a hybrid *Aa*, and the tendency of the population derived from the cross to revert to the two parental forms *AA* and *aa*. He showed that in the nth generation of self-fertilization the proportion of genotypes would be $(2^n-1)AA : 2Aa : (2^n-1)aa$. This was the first generalization predicting the genetic constitution of a population under a specified system of mating. The equilibrium relation $q^2AA : 2q(1-q)Aa : (1-q)^2aa$, often referred to as the Hardy-Weinberg equilibrium expression, is an application of the same principle to a cross-breeding population mating at random.

Mendel's generalization implied the latter one because it assumed that automatic consequences follow from the operation of the segregation principle and the integrity of the hereditary elements. The consequences will thus differ depending upon the prevailing system of mating. Close inbreeding will result in a decrease in the frequency of heterozygotes and in differentiation among pure lines while the proportions of genotypes will be maintained under random mating. Castle (1903) and Pearson (1904) grasped the second idea; Hardy

(1908) and Weinberg (1908) generalized this equilibrium idea to apply to any array of gene frequencies at any single locus, although this seems to have been at least implicit in Mendel's paper and in Pearson's (1904) extension of it. Chetverikov (1926), who initiated the modern phase of experimental population genetics, based his ideas on "Pearson's Law of Stabilizing Crossing"—a clear recognition of the conditions of stable equilibrium.

G. H. Hardy, professor of mathematics at Cambridge University, was stimulated by R. C. Punnett to publish his derivation of the equilibrium formula. Punnett reported to Hardy that G. U. Yule, the statistician, had suggested, as a criticism of Mendel's principle, that a dominant gene, once it had entered a human population, should produce a distribution of $3/4$ of the dominant phenotype and $1/4$ of the recessive (cf. Punnett, 1950). Hardy, referring to Pearson's proof of the stability of the 1 : 2 : 1 ratio then derived the constant ratio of the three genotypes pAA : $2qAa$: raa as $q^2 = pr$, whatever the values of p, q and r. This is now usually stated in the form $q^2AA + 2q(1-q)Aa + (1-q)^2aa = 1$.

However, Weinberg (1908), a practicing physician of Stuttgart, had already worked out the consequences of Mendelian heredity in human populations. He had learned about Mendelism in 1905 and began then the theoretical work which led him to generalize the equilibrium principle to include the existence of multiple alleles at the same locus. Later he worked out the manner in which mutations at different loci, with independent assortment, come to equilibrium under continued random mating in large populations. Weinberg's three papers of 1908, 1909, and 1910 established him as one of the founders of population genetics. This opinion, expressed by Curt Stern (1962), in a biographical note about Weinberg, was reached belatedly by many geneticists who had not studied Weinberg's papers before his death in 1937. His study, in 1909 and 1910, of the correlations between close relatives dealt not only with genic but environmental variance and methods of partitioning such variances. In this he anticipated some of the main results obtained independently ten years later by R. A. Fisher (1918). In 1912, he developed the methods of correcting expectations for Mendelian segregation from human pedigree data under different kinds of ascertainment applied to data from small families. These and a variety of other methods used in the study of human genetics appeared later in the works of Felix Bernstein, Fritz Lenz, Gunnar Dahlberg, Lancelot Hogben and others. Weinberg's highly original work failed to interest his contemporaries, who were not prepared to find in algebra and probability theory the keys to the fields of population genetics and human genetics.

Mendelian equilibrium as a consequence of segregation of alleles under random mating was nevertheless probably rather widely recognized in a general way before 1910. For example, W. J. Spillman, who was himself responsible for advances in genetics (*cf.* p. 85) as well as the rapidity with which Mendelism influenced agricultural practice in the United States, said in an agricultural publication in 1909:

> If corn were completely cross-fertilized, the proportion of these 9 types (F_2 genotypes) would be approximately the same the second year and each year thereafter, assuming of course that all types are equally productive.

The idea had already appeared in the work of Shull and of East (*cf.* p. 124).

Although the chief theoretical work on a second problem opened up by Mendel's generalization, *viz.*, systems of mating, was not taken up until after 1910, an important influence in the earlier period was Johannsen's work with pure lines. Johannsen was stimulated by Galton's predictive model based on blending rather than particulate inheritance. The model dealt with random-mating populations and served as the stimulus for Johannsen's experimental analysis of populations. In Johannsen's theory, "populations" of self-fertilizing organisms were mixtures of genotypes to be contrasted with "pure lines," *i.e.* the progeny descended by self-fertilization from single individuals and therefore propagating constant genotypes. Under random mating, with Galton's assumptions, populations could not for long have retained their mixed character but were destined to become uniform without inbreeding. Johannsen's theory, based on retention of integrity of individual genes, predicted the mixed character actually found in cross-fertilized populations, such as those of maize, and the resolution of a heterozygous population into a series of more or less constant pure lines as the result of artificial inbreeding. The demonstration of this predicted outcome by Shull (1908) and East (1907) represented an important step in the development of population genetics.

Heterosis, the concept of the superiority of heterozygotes, was introduced by G. H. Shull in 1908 although the word itself was first used in 1914 (Shull, 1952). Heterosis has since played an important role in the interpretation of genetic polymorphism and of the maintenance of variety in wild and domesticated populations in addition to its primary effects on the theory and practice of plant and animal breeding. These papers of East and Shull led the way toward the experimental establishment of the view that the outcome of different systems of mating is determined by the operation of the Mendelian mechanism (East and Jones, 1919).

Some Applications of Mendelism

Distinctions between theoretical and practical, pure and applied, have never been sharply drawn in genetics. From the beginning, this has been in part responsible for the rapid progress of genetics. Two cases from the first years of genetics illustrate this feature.

The first has to do with disease resistance in plants. At a time when the evidence for Mendelian heredity came largely from morphological and color variations of the kind dealt with by Mendel, R. H. Biffen of Cambridge University published clear evidence (1905) that resistance to yellow rust in wheat was inherited as a simple Mendelian recessive to susceptibility. This made it possible to derive, in the F_3, families which breed true to resistance but which contain other desirable traits derived from a susceptible line. It should be recalled that Correns (1900) had discovered in the four o'clock a gene determining a localized disease. Its normal allele might thus be said to be the first gene identified by disease resistance. Eventually the rationale and the speed of improvement of crop plants were both greatly affected by this discovery; and the inheritance of disease resistance became one of the brightest chapters in plant breeding.

It also foreshadowed the discovery of the kind of lock-and-key relationship between the genotype of the host and the genotype of the parasite which led to a view of the essentially physiological or biochemical nature of the change wrought in development by a changed gene.

The relations of genetics and plant pathology have been well reviewed by J. C. Walker (1951).

The second example also stems from work begun in the first decade of Mendelism which resulted in what has been called the greatest success story of genetics—hybrid corn. This began with an attempt by Shull (1908) to analyze the inheritance of quantitative characters. Influenced by Johannsen's pure-line concept, he inbred a number of lines of maize (normally a wind-pollinated plant) by artificial hand pollination in order to reduce their genotypic variability and to fix certain traits as homozygous. He then crossed inbred lines differing in numbers of rows of seeds on the ear. He found that all aspects of size, vigor, and productivity declined as inbreeding proceeded; but when crosses were made between members of different inbred lines the F_1 offspring showed a return to at least the original vigor and size of the original parents before inbreeding began, and sometimes even an increase over that. Similar results were obtained by East, beginning in 1905 but not available in written form until 1912 (East and Hayes, 1912). Shull (1908) pointed out that the decrease in size and vigor paralleled exactly the rate at which homozygosis was attained. He supposed a field of maize to be a collection of hybrid genotypes which owed their vigor to their heterozygous nature. By inbreeding, these could be separated into more homozygous, and weaker, genotypes. In 1909, he proposed a method of breeding which eventually (when combined with ideas from East, Hayes, and Jones) gave rise to the hybrid corn breeding program which caused a major revolution in American agriculture (Manglesdorf, 1951; Jones, 1959). The essence of it was to maintain inbred lines of maize solely for the purpose of utilizing the hybrid vigor or heterosis resulting from crossing different lines. A decisive suggestion which made hybrid corn practicable and profitable was that of D. F. Jones who introduced the method of crossing two different F_1 hybrids ($A \times B$; $C \times D$; $F_1 A B \times F_1 C D$) and thus taking advantage of the large yield of high quality seeds on the F_1 plants. The whole method was based on a Mendelian interpretation of the effects of inbreeding, later set forth in detail in East and Jones (1919).

This account of developments in the period 1900–1910 cannot be concluded without mention of the discovery by Karl Landsteiner, during 1900 (*cf.* Landsteiner, 1901) that human blood can be classified into groups according to reciprocal relations between agglutinability of red-cells and agglutinating capacity of blood serum. It was only at the end of this decade that von Dungern and Hirszfeld proved that the individual differences in the red cell antigens (the ABO system) which Landsteiner had discovered were determined exclusively by the genotype. Even though the details of the interpretation were not correct, this was one of the origins of exact knowledge of heredity in man which, together with Garrod's discovery of the genetic basis of several metabolic abnormalities, could have led in this decade to the development of human genetics. However, the development was delayed for another twenty years.

Summary

In reviewing, however incompletely, the history of the first decade of Mendelism, I have been impressed by the fact that many of the steps which proved to be essential for later progress were taken early in this period. It is evident that in

1900 the time was ripe for a rapid development of ideas about heredity. De Vries and Correns, at least, had already begun in the early 1890s the same kind of experiments which Mendel had carried out. Stubbe has pointed out (1963, p. 201) that the plant breeders Wilhelm Rimpau in Germany in 1891 and P. Bolin and H. Tedin at Svalöf in Sweden in 1897 had obtained results from hybridization experiments, begun in 1890, which could readily be interpreted on Mendel's principles. The work of Spillman (1901) at the Washington State Experiment Station, U.S.A., has already been mentioned, and Roberts (1929) has listed several other botanists engaged upon hybridization work before 1900. Similar experiments with animals had been undertaken before 1900 by Haacke, von Guaita and others. Although these did not yield any rules comparable to Mendel's, they were indications of the interests of the times. Bateson had begun his breeding experiments with poultry before 1900 and had results ready for a Mendelian interpretation soon after he found Mendel's paper.

The rapidity with which the cytological basis of Mendelism came to view, within the first two or three years after the rediscovery, shows that cytology was ready for new interpretations. Its progress was stimulated by experimental genetics and it in turn enhanced interest in heredity.

Probably the chief causes of the burst of interest and research activity in genetics in the 1900–1909 period are found in the characters of the persons responsible for it. Scientists of the intellectual quality and vigor of de Vries, Correns, Bateson, Boveri, Sutton, Wilson, Castle, Cuénot, and Morgan are rare in any field, but that they should concentrate their efforts on one set of problems at one time is an unusual event. It rapidly produced an unusual result in the creation of a new science.

Part III

THE THEORY OF THE GENE: 1910–1939

– Chapter 13 –

GENETICS IN 1939: MILEPOSTS OF PROGRESS

THE MAIN LINE of advance in the period 1900–1909 was
the confirmation, consolidation, and extension of Mendelism.
After 1910, the center of the stage was occupied for the next
twenty-five years by what began as the chromosome theory (in
Sutton's sense) and turned into the theory of the gene, as
described by Morgan and his fellow workers. During this period also the increasing number of persons devoting their research efforts to genetics and the spread of their interests to
new areas resulted in a great diversification of genetics. Its
essential unity was, however, as clear as ever, retained because
of the recognition of the gene as both the conceptual and
physical element by means of which order could be perceived
in the processes of heredity and mutation. The gene could
well give its name to a period in which all of the problems
of classical genetics came to maturity.

The central theme throughout the period was analysis of
the material basis of heredity in higher plants and animals.
The analysis was carried out by experimental breeding (genetic
analysis), by cytological observations, by the experimental
study of gene mutation and chromosome changes, aided by
radiations of various kinds, and, toward the end of the period,

by more direct chemical and physical study of chromosomes and their constituent proteins and nucleic acids.

This concern with the physical substratum of heredity and variation was in marked contrast to the abstract or statistical conception of the gene which Mendel had introduced and which had been so strongly supported by Correns, Bateson, and Johannsen. These pioneer geneticists warned of the confining effects of any corpuscular theory. In fact, none of them participated in the development of the chromosome or gene theory, and Johannsen (1923) criticized it severely. Other biologists, not primarily concerned with genetics, also suspected the trap of preformation concealed within the theory of the gene—and considered that the orderly structure which arose in the ten years 1910–1920 might collapse from the very weight of pure mechanism which was used in building it (Russell, 1930). Such fears had some justification, for Mendelism itself had only recently displaced views derived deductively from the material particles on which Spencer, Darwin, Galton, and Naegeli had based their now-discredited theories of heredity. Was the pendulum now going to swing back to nineteenth-century ideas? As it turned out, the bugbear of preformation simply had to be lived with until the nature of genes and chromosomes could be viewed in dynamic terms rather than in the static ones of preformism.

The strongest influence in forming such conceptions of the gene came from experimental studies of gene mutation and especially from the proof that this process, which was shown to have a low rate of "spontaneous" occurrence, could be enormously speeded up by radiations of various kinds. The idea that change in the gene could be caused by its being hit by a particle, or by its ionization path (the Treffer theory), was certainly a strong reason for regarding the gene itself as a particle. Since it retained its form through many generations, yet was changeable, the problems of constancy and variability in organisms could be viewed at the genic and eventually at the molecular level and thus studied experimentally by chemical and physical methods.

Although the paradox remained to which Morgan had called attention in 1910—unlikeness in form and function among the cells of the individual combined with identity of their chromosomes and genes—efforts to resolve it began in earnest during this period. A new field appeared to which Valentin Haecker in 1918 gave the name "phenogenetics." It was intended to elucidate the manner in which genes controlled development.

It proposed to trace differences in development associated with genic differences such as AA and aa and thus could deal with details of structure or color which had been shown to "Mendelize." The results were thus largely descriptive and confined to plants and higher animals in which such gene differences had been identified.

However, ideas of greater analytical usefulness, since they led in the direction of a general theory, had appeared soon after 1900 when genes were assumed to make enzymes (Cuénot, 1903; cf. Haldane, 1954, p. 16). That genes control enzyme specificity appeared highly probable even before specific enzymes were found to be controlled by specific genes.

Another contribution of genetics to physiology took form in this period. It arose out of attempts to localize genes in chromosomes and to analyze chromosomal aberrations. The balance theory of sex led to a general theory of the manner in which genes control the balance of reaction rates.

Near the end of the period, studies began which resulted in a general outline of the manner of genetic control of metabolism. The discovery of the "position effect" related the functioning of the genetic materials in development to their locations in the chromosomes.

The theories of population genetics underwent their basic development in this period and a beginning was made in testing them experimentally. This made it possible to conceive of the causes of evolution as agencies affecting the frequencies of genes in populations: pressures of mutation, selection, migration, and chance or random variations dependent on population size. There was intense activity in the genetical study

of species differences both in respect to their genic and chromosomal complements and in extra-chromosomal structures
(plasmon or plastone), all of this especially in plants. In
animals this work was facilitated by the exploitation of the
newly rediscovered giant salivary chromosomes in certain dip-
terans, especially Drosophila species.

Mileposts of Progress

The decade preceding the outbreak of the Second World
War was the climactic one of classical genetics. A state had
been reached in the study of each of the main questions from
which the outline of future study could be clearly seen: the
nature of the transmission mechanism and of the gene, the
causes of mutation and of evolution, the manner in which
genes control metabolic processes. This progress produced a
spirit of confidence and optimism which colored the research
reports and textbooks at the time.

In this same decade, however, influences acting in the opposite direction, against progress in genetics, made their appearance in Germany and the USSR and spread to countries
which fell under the influence of these two great powers. The
roots of political suppression of genetics in the USSR are
traceable to the early 1930s, but the main effects were not
apparent until the outbreak of the war. Then genetics was
deprived, for at least a generation, of an important source of
ideas and of theoretical and experimental research. In Germany, beginning in the early 1930s, events took a different
course and arose from different causes, but the effects were
similar—the loss for many years of the potential contributions
of German genetics which had formerly provided leadership
in the field.

Not the least of the effects of these two disasters for genetics
was the realization that the soundness of its scientific foundation was not by itself a sufficient condition of progress. These
harsh reminders of the dangers to which the science was to
be exposed may well have moderated the confidence of many

geneticists. This would however be less likely to affect those who in the following period were to devote themselves to studying genetics at the molecular level.

One can gain perspective for viewing the period beginning in 1910 by considering some of the important publications which appeared in 1937–1939. These marked the state to which certain key questions had attained, and in some cases pointed the direction of future developments. A few such have been chosen as mileposts of progress and as evidence that the choice of the last years of the 1930s as a cutoff point is not entirely arbitrary.

As a gauge of the union of genetics and cytology, Darlington's *Recent Advances in Cytology* appeared in a second edition in 1937. The first edition (1932) had been the most influential book in cytology since Wilson's *The Cell in Development and Heredity*. It radiated confidence in the ability of the new combination of cytology with genetics to answer all of the questions about protoplasm. As J. B. S. Haldane pointed out in a foreword to the second edition:

> It is perfectly possible that *Recent Advances in Cytology* marks a turning point in the history of biology. For some centuries the deductive method in the biological sciences has been very properly suspect. But first in genetics and now in cytology, we have returned to it. General principles have been discovered of such wide validity that we can predict from them with considerable confidence, and on the rare occasions when the prediction is falsified, we are inclined to look for undetected causal agencies rather than to recast our first principles. This attitude has long been normal in chemistry and physics. Its introduction into biology, however unwelcome it may be to conservative biologists, is a sign of the growing unity of science.

This was, of course, an overstatement, and progress continued to depend on experiments and inductively derived conclusions, but it was true that certain principles had attained what appeared to be a universal validity. One of these was that chromosomes were made up of linear successions of different genes and behaved as they did because of this fact. Genetically ascertained facts were in the future to be used to explain cytological ones and not, as had originally been proposed, the other way around.

Hans Bauer's review of progress in cytogenetics which appeared in the *Fortschritte der Zoologie* in 1937 was another sign of the times. All references were to work carried out with Drosophila. Drosophila, several species of it in addition to the classical melanogaster—dominated and almost monopolized animal cytogenetics, although pure cytology dealt with a wide variety of species.

The plant which appeared to have contributed essential proofs linking individual loci with chromosome segments and behavior was maize, and this progress was reviewed by Marcus Rhoades and Barbara McClintock (1935).

Theodosius Dobzhansky's book of 1937, *Genetics and the Origin of Species,* was a landmark, indicating attainment of a new stage in restating the problems of evolution in terms of modern cytogenetics and opening a new era of evolutionary biology.

A review of genetics for students, including literature through 1937 (Sturtevant and Beadle, 1939), devoted about two-thirds of its 380 pages to the chromosome theory and developments which had grown out of it. In the same year there appeared Darlington's *Evolution of Genetic Systems.* The subject was genetics, but based entirely on the chromosome theory.

In 1938, Richard Goldschmidt's *Physiological Genetics* showed how little essential progress had been made by the phenogenetic methods of tracing the developmental history of differences due to one or a few mutant genes. The chief sign of the times in this book, however, was its attack on the theory of the gene. On page 303, Goldschmidt asked a question that would be pertinent for many years: "Is not the whole conception of the gene as a hereditary unit obsolete?" He based his question on the demonstration, in a few cases, of an inherited change in phenotype due to a rearrangement of genic material not involving gene mutation—the so-called posi-

Gregor Johann Mendel
(1822-1884)

Mendel's experimental garden plot.

Hugo de Vries
(1848-1935)

Carl Erich Correns
(1864-1933)

Erich von Tschermak-Seysenegg
(1871-1962)

Wilhelm Ludwig Johannsen
(1857-1927)

August Weismann
(1834-1914)

Walter Stanborough Sutton
(1877-1916)

Theodor Boveri
(1862-1915)

William Bateson
(1861-1926)

Archibald Edward Garrod
(1858-1936)

Wilhelm Weinberg
(1862-1937)

Edmund Beecher Wilson
(1856-1939)

Thomas Hunt Morgan
(1866-1944)
and
Rollins A. Emerson
(1873-1947)

Calvin Blackman Bridges
(1889-1938)

tion effect, of which Bar eye in Drosophila was the first example. This means, he said, that the chromosome itself is:

the actual hereditary unit controlling the development of the wild type, that purely steric changes at the individual points of its length produce deviations from wild type which may be described as mutations, even as point mutations, though no actual wild type allelomorph and therefore no gene exists.

George W. Beadle's review in the first volume of the *Annual Review of Physiology* (1939) was based mainly on work published in 1937 and 1938. It pointed to the new trend in the utilization of gene-controlled differences in revealing sequences of chemical reactions leading to the synthesis of plant and animal pigments, especially the synthesis of ommatins in insects as studied by the newly developed methods of tissue transplantation. In referring to a paper by T. Caspersson and J. Schultz (*Nature*, 1938) he commented: "It is also suggested that there is a relation between nucleic acid and gene reproduction." The review leaves the impression that genetics was moving in the direction of physiology and biochemistry. Beadle disagreed with Goldschmidt's attack on the gene and found "the views of Goldschmidt difficult to reconcile with many of the well-established facts of modern genetics." The new genetics (in Beadle's view) would thus be erected on the basis established by the theory of the gene.

At about the same time (1939) appeared an imaginative essay by Nikolai Koltzoff, the director of the Institute of Experimental Biology in Moscow. This institute had become a center of progressive work and thought in what was then in fact "the new genetics," although this name was later claimed for the obscurantist jumble of philosophy and "agro-technics" which went by the name of Lysenkoism. Koltzoff's essay "Les Molecules Héréditaires" was at the opposite pole from the Lysenko trend. It took the results of the rapid progress in cytogenetics and extended it by the logical next step, asking what kind of molecule composed the gene string ("genonema," a new term for chromosome). He could point to the converging research of physical chemists and of geneticists, working apart from each other, which centered on the idea of a long polypeptide chain with billions of isomeric parts.

Doubtless other evidence could be cited that the end of the thirties was a period of climax and impending change, the beginning of the period of transition from classical, or formal genetics, to molecular genetics as it exists today. It seems a good vantage point from which to view the changes since 1910.

– Chapter 14 –

THE THEORY OF THE GENE

THE CHROMOSOME THEORY was based mainly on the re-markable animal *Drosophila melanogaster*, the small dipteran sometimes miscalled a "fruit fly" but correctly identified as the "pomace fly", or "vinegar fly." Its great advantage was rapidity and ease of cultivation. A single pair could, in less than two weeks, provide several hundred offspring which could be observed at any stage: egg, three larval instars, pupae, and imagines, or adults. Phenotypic differences in morphology, pigment, viability, behavior, and many physio-logical traits were readily analyzed in terms of segregating alleles by simple experimental breeding methods.

The first mutant to be analyzed, the white-eyed one re-ported by Morgan in 1910 as sex-linked (he first called it "sex-limited") was soon followed by another sex-linked re-cessive, rudimentary wings, and a little later yellow body color turned out to be sex-linked and recessive to the normal grey body. The factors for white eye and for yellow body tended to be inherited in the same associations in which they entered the cross if both came from the same parent—*i.e.*, from a cross of yellow-white by grey-red, they tended to be coupled, usually appearing together in the F_2 but with a few flies showing yellow bodies with red eyes and grey bodies with white eyes.

This was in marked contrast to what would be expected if Mendel's second rule applied. From these and similar results with other sex-linked traits, Morgan concluded (1914, pp. 92–93):

> We may make a general statement or hypothesis that covers cases like these, and in fact all cases where linkage occurs: viz. that when factors lie in different chromosomes, they assort freely and give the Mendelian expectation; but when factors lie in the same chromosome they may be said to be linked and they give departures from the Mendelian ratios. The extent to which they depart from expectation will vary with different factors. I have suggested that the departures may be interpreted as the distance between the factors in question.

At this point (winter, 1910–11), Morgan took two Columbia College undergraduates to work in his laboratory, Alfred H. Sturtevant and Calvin B. Bridges. This was the beginning of the famous "fly-room" on the sixth floor of Schermerhorn Hall in the center of the Department of Zoology and near the laboratory of E. B. Wilson. Another early member of the group was Herman J. Muller, also a Columbia College undergraduate who entered graduate work as a student of E. B. Wilson. Later the small fly-room had eight desks crowded into it along with incubators, racks of fly bottles, etc., and here in close proximity worked Morgan, Bridges, Sturtevant, students and others engaged in "fly-work," and a changing company of postdoctoral students from other countries, beginning in 1917 with Otto L. Mohr from Norway.

I stood in the doorway of that room in the spring of 1914, having read Morgan's *Heredity and Sex* which had inspired in me the hope that I might be allowed to undertake graduate work with the group in the following year. But as Morgan explained, and as was obvious from the doorway, there was little room; also, no one knew of a means for supporting another graduate student. Fourteen years later, in 1928, when I had joined the zoology department, it fell to my lot to prepare the "fly-room" for other uses after the departure of

Morgan and his associates for the California Institute of Technology. The ingenuity with which simple homemade installations had been made to serve as the basis for a major scientific advance then became apparent to me.

An account of this period has been given by one of the participants (Sturtevant, 1959, p. 295) :

> This group worked as a unit. Each carried on his own experiments, but each knew exactly what the others were doing, and each new result was freely discussed. There was little attention paid to priority or to the source of new ideas or new interpretations. What mattered was to get ahead with the work. There was much to be done; there were many new ideas to be tested, and many new experimental techniques to be developed. There can have been few times and places in scientific laboratories with such an atmosphere of excitement and with such a record of sustained enthusiasm. This was due in large part to Morgan's own attitude, compounded of enthusiasm combined with a strong critical sense, generosity, openmindedness, and a remarkable sense of humor. No small part of the success of the undertaking was due also to Wilson's unfailing support and appreciation of the work—a matter of importance partly because he was head of the department.

In the two years following his first Drosophila paper of July, 1910, Morgan published thirteen papers on the occurrence and behavior of some twenty sex-linked mutants in Drosophila. The basic theory of linkage as due to location of the linked genes in the same chromosome, and breaks in the linkage as due to crossing over according to the chiasmatype theory of Janssens, appeared in September, 1911, although Morgan (1911, p. 384) had first stated in a public lecture on July 7, 1911, at Woods Hole, Massachusetts, that:

> during segregation certain factors are more likely to remain together than to separate, not because of any attraction between them, but because they lie near together in the chromosome, as will be explained more fully later.

By November, 1912, linkage of non-sex-linked genes had been found, as well as the failure of crossing over to separate such linked genes in the male. The theory of the mechanism of Mendelian heredity was now in full cry, and was to retain the same feverish pace of development for another fifteen years.

In reading this first flood of papers, the first grist from the fly-mill, one is struck by the sudden change in attitude on Morgan's part toward Mendelism and toward the chromosome theory. Morgan's skepticism concerning the integrity of "unit factors" and of "purity of the gametes," two cornerstones of Mendelism, probably accounted for his entrance into genetics in the first place. His early papers on inheritance in mice and rats (written from 1904 to 1906) were chiefly concerned with pointing out departures from results expected on the basis of Mendelian theory.

It may provide some amusement for those who first became familiar with Morgan's theory of the gene in its definitive form to learn that, in 1909, Morgan entertained no high hopes for the chromosome theory. His long critical paper, "Chromosomes and Heredity," was published in August, 1910, but it must have been written at least a year earlier. It appeared *after* the paper reporting the first sex-linked mutant in Drosophila which was dated July 7 and appeared in print July 22, 1910.

It is mainly directed at explaining development, always Morgan's primary interest. After discussing the untenable Roux-Weismann idea of mitoses as qualitative and Boveri's disproof of it, he faces up to the chief question, "Chromosomes and Mendelism." Sutton (1903), Wilson (1903), and Boveri (1904) had made it apparent, Morgan said, that synapsis is the moment of segregation. (He might have noted that Correns' paper of 1902 had worked this out in detail, including a mechanism of crossing over.) Morgan stated that the trouble was that there were many more characters than chromosomes, hence there must be several characters per chromosome and these "must Mendelize together. But they do not." He found in the absence of grouping of characters a fatal objection to the chromosome theory. Moreover:

> If Mendelian characters are due to the presence or absence of a specific chromosome, *as Sutton's hypothesis assumes,* how can we account for the fact that the tissues and organs of an animal differ from each other when all contain the same chromosome complex?

Fifty-five years later, this is still the unresolved paradox, but we no longer find in it an insuperable objection to the chromosome theory as a view of the transmission mechanism of heredity. Within a year from the time that Morgan uttered these prophecies of doom for the new chromosome theory, he had so far forgotten them as to have put aside, temporarily it is true, that part of the problem which dealt with development, and had concentrated his efforts on the analysis of what he soon came to call "the mechanism of Mendelian heredity." Morgan was well aware of the objections which would be raised against the chromosome theory, since he had entertained many of them himself.

It was Morgan's skeptical, critical attitude which accounted not only for his first entrance into genetics but also for the solid experimental base that he laid for the new form of the chromosome theory. He did not accept, at least for very long, any idea that could not be put fairly quickly to experimental test. His very impatience was one of the reasons for his success.

With all his enthusiasm for the new experimental animal and the general theory that was quickly forming from the new work, the problems of development were never really abandoned. All through Morgan's "fly-years" there appeared papers testifying to his interest in regeneration, sex-determination and differentiation, self-sterility and other problems in a great variety of animals. Sturtevant (1959) compiled a list of fifty kinds of animals on which Morgan published results of investigation. It is evident, however, that from 1910–1920 the

experimental breeding work with Drosophila occupied the
center of the stage for him and represented that degree of
liberation, sometimes grudgingly accepted on his part, from
the ultimate questions of organization and integration for
which no practical means of study could then be found.

The Drosophila material was well adapted to his needs. He
bred his flies in milk bottles with bananas as food and with
used envelopes from his correspondence as supports on which
the larvae could pupate. He classified the flies (under ether)
with a hand lens or even without one. Increasingly, his
younger associates expanded the breeding operations and un-
dertook the cytological observations which became increasingly important.

In 1913, Sturtevant produced the first chromosome "map"
by showing that six sex-linked mutants could be arranged, on
the basis of crossing-over frequencies, in linear order in the
X chromosome, and other maps of other chromosomes followed based on the principle that "map distance" was proportional, at least for closely linked genes, to the percentage
of recombination (crossing over) between them.

This paper of Sturtevant's provided the first test of Morgan's basic hypothesis that the frequency with which linked
genes separate when the heterozygote forms its gametes is a
measure of the relative distance between them. Morgan had
taken the idea of crossing over, as the physical counterpart
of the separation or recombination of adjacent genes, from
Janssens' chiasmatype theory (1909), assuming that when
pairs of homologous chromosomes twist about each other in
synapsis they exchange parts, or "cross over," at the chiasma.
When they then split apart in a single plane, the original
order of materials in each homologue will over short distances
be more likely to be retained, while more distant regions run
a greater chance of being separated by a chiasma. Sturtevant's
observations tested one of the consequences of this view—
namely, that linked genes should be in a single linear succession—without, of course, bearing on the question whether a
crossover in fact coincides with a chiasma. The order of the

six loci marked by mutant genes was proved to be linear when linkage strength was measured by proportion of crossovers, but crossing over gave a good estimate of distance only when the proportion was low, meaning when the loci were near. More distant loci such as A and C were found to retain their original associations by being brought together after being separated by one crossing over between A and B and by the occurrence of another between A and C. Thus, in a hetero-zygote $\frac{ABC}{abc}$ gametes were shown to occur as follows:

ABC	noncrossover
abc	noncrossover
ABc	crossover between B and C
abC	crossover between B and C
Abc	crossover between A and B
aBC	crossover between A and B
AbC	double crossover: A and B, B and C
aBc	double crossover: A and B, B and C

The apparent "distance" between A and C would thus appear to be less than the sum of AB and BC. Sturtevant worked out the corrections required by the occurrence of double crossing over, constructed the first map by using closely linked genes only, and in general laid out the chief methods which underlie mapping in Drosophila and other organisms.

Crossing Over

At about the same time, Muller, in a series of four papers published in 1916, subjected the mechanism of crossing over and the basic theory of the linear order of genes to a searching theoretical and experimental study. One object was to test Janssens' chiasmatype theory, which, as Muller said, Janssens had "intended to explain the supposed fact that there might be more pairs of factors capable of recombination than pairs of chromosomes." Morgan had invoked the chiasmatype hypothesis for a more specific purpose, to explain recombination of linked factors. Muller concluded from his examination of then existing evidence that establishment of a linear order by crossing-over frequencies would in fact prove that an actual interchange of blocks of genes, segments of the map, occurs at meiosis. If in fact the interchange occurs at synapsis, than it would constitute crossing over. The connection between interchange and meiosis was shown by Bridges' new evidence from nondisjunction, in which X chromosomes which had paired with Y showed no crossing over while those which paired with X did.

Secondly, Muller pointed out that the total amount of crossing over in each of the four chromosomes of Drosophila, giving its "map length," corresponded roughly with the relative cytological length of the chromosomes. The telling argument was the discovery by Muller (1914) of a gene which, since it was linked to no other group, was ascribed to the very small fourth chromosome. This was followed by a second mutant gene with similar behavior (Hoge, 1915). In accordance with a prediction based on the small physical size of the fourth chromosome, these two genes showed no crossing over and in general complete linkage was found to be the rule for genes in this chromosome. Thus, relative sizes of chromosomes were in harmony with assumptions called for by the chiasmatype theory of crossing over.

Muller's chief evidence was derived from a theoretical and experimental study of the interference of one crossover with another in the same chromosome. He constructed special stocks of Drosophila by adding twelve mutant genes one at a time to serve as markers in the X chromosome, and ten in chromosome II. Extensive tests with these showed that crossing over in one chromosome had no effect on crossing over in the other, but that within one chromosome, crossing over in one region interferes with crossing over in adjacent regions, and that this influence declines with distance and disappears at forty to fifty units of crossover distance. If the genetic map length is of the order of 100 and crossing over depends on interchange at points of crossing (chiasmata), then there should be about two loops between chiasmata per chromosome. This seemed to Muller to exclude crossing over at later stages in meiosis, since Janssens' (as well as others') cytological observations showed many twisted loops at those stages. Crossing over should therefore occur in an early stage in synapsis; Harold H. Plough (1917) soon showed that crossing over in Drosophila females could be influenced by altering the temperature, but only at the time when oöcytes were at an early synaptic stage.

Muller's study was important in showing how far theory could be carried by purely genetical methods, and this became one of the hallmarks of the work of the members of the Drosophila group. The meiotic stages of Drosophila were intractable cytologically—there was an interval of fifteen years before Stern finally proved directly, in 1931, that genetical crossing over was accompanied by interchange of actual segments of homologous chromosomes. However, so thorough had been the genetical experiments, like those of Muller, that Stern's demonstration seemed anticlimactic. Faith in its basic assumptions had already guided the analysis of the genetical material for twenty years.

These studies of linkage and crossing over involved no actual test of the basic assumption that genes are in fact parts of chromosomes. Strict proof of this by a novel method began to be provided by Bridges' discovery in 1913 of nondisjunction of the X chromosome and later through analysis of this by both cytological and genetical methods. But before that essential step in the development of the chromosome theory can be described, other work, chiefly by Wilson and his students, must be referred to, for it was this which laid the cytological foundations of the theory of the gene.

− Chapter 15 −

THE CYTOLOGICAL BASIS
OF THE CHROMOSOME THEORY

THE PREPARATORY PROOFS had begun with Henking's demonstration (1891) that the reduction division begins by conjugation of chromosomes, two by two, and with J. Ruckert's suggestion (1892) that the conjugants had come one from each parent, and that they could exchange material and thus produce chromosomes with parental characters in new combinations. Montgomery (1901) and Sutton (1902, 1903) built on this ground in supposing that the mates in synapsis were always one paternal and one maternal chromosome of the same pair, and that the maternal and paternal members could go to either pole of the mitotic figure irrespective of the positions of other pairs, thus producing independent assortment of maternal and paternal chromosomes and a mechanism for the independent assortment of characters which Mendel had found. This meant essentially that, if genes are in chromosomes, the number of groups of linked genes should be the same as the number of pairs of chromosomes—an assumption later proved by Morgan and his co-workers for Drosophila.

But as McClung found (1902), confirming Henking, and as was later worked out by Wilson and others, not all chromosome pairs in animals consist of equal and like numbers. Some pairs consist of visibly different members: they are "heteromorphic." The first case of marked heteromorphism was found in the sex chromosomes in certain insects, where females had an equal pair XX and males a single X, a condition referred to as XO. This appeared at first to be the case in Drosophila; but in 1908, Nettie M. Stevens showed that in the species first studied by Morgan and his group, *Drosophila melanogaster* (*ampelophila* in the earlier literature), the female has four pairs of chromosomes, with pair members alike, or homomorphic, while in the male three of these pairs, the autosomes, are like those in the female, but one pair concerned with sex is markedly heteromorphic, XX in the female, XY in the male. At the same time that Bridges was utilizing the work of Stevens and of Wilson in identifying the chromosomes of Drosophila, other investigators, especially Eleanor Carothers (1913), were using differences between members of homologous pairs of chromosomes to give direct proof of the independent assortment of members of different pairs—which Sutton had assumed as a consequence of the location of Mendelian factors in chromosomes. Carothers found in the same insect studied by Sutton, the lubber grasshopper *Brachystola magna*, that other pairs of chromosomes might be morphologically heteromorphic. In one such case she counted 300 spermatocytes in which one member of an autosome pair (the larger one) went to the same pole as the X chromosome in fifty-one percent of the number observed, while in forty-nine percent it went to the other pole. This evidence of independent assortment of members of different pairs of chromosomes was extended by Carothers and Robertson and Wenrich to so many other animals as to leave no doubt that it constituted the general rule anticipated by Sutton: chromosome pairs behave in meiosis like pairs of alleles.

But to return to Bridges. His doctoral dissertation was published (1916) as the first paper in the first volume of the first American journal devoted exclusively to genetics. It had the title "Nondisjunction as Proof of the Chromosome Theory of Heredity." It contained evidence that anomalous transmission of a sex-linked mutant gene was caused by the anomaly in the transmission of the sex chromosome which had been predicted from the genetical evidence and then verified cytologically. This was the first verified case of nondisjunction. Females with a sex-linked recessive, expected to produce male offspring, all of which would inherit the mother's X chromosome and show recessive characters carried by it (since the Y chromosome from the father transmits no normal alleles to prevent the expression of genes in the X), actually produced some males which were normal. Such females proved to be XXY; their cells had nine rather than the normal eight chromosomes, that is: two pairs of V-shaped chromosomes (chromosome pairs II and III); one pair of small dot-like chromosomes (pair IV); one pair of rod-shaped telocentric ones (the X chromosomes), and one with a hook, recognizable as the Y chromosome. The eggs from such a female would all contain one of each of the autosomes, but there would be four kinds of eggs with respect to sex chromosomes: XX, XY, X, and Y. The Y egg fertilized by an X sperm from a normal male would produce a male inheriting sex-linked genes from the father rather than from the mother as usual. Thus an X^wX^wY (white-eyed) by a wild-type red-eyed male X^+Y would give X^+Y (*viz.* red-eyed males) as exceptions due to nondisjunction, *i.e.* failure of the two X chromosomes to disjoin and enter separate eggs, thus producing some eggs XX and others Y. These exceptions, and others due to the same process, can be explained *only* if the sex-linked alleles w and + and other sex-linked alleles are always transmitted in the X chromosome, and if the Y chromosome carries no alleles for these.

This phenomenon, which was proved for X chromosomes marked by one of several sex-linked mutants (and later for the small fourth chromosome) was discovered as soon as sex-linked genes were found. Morgan and Bridges at least suspected its interpretation even before full cytological proof was obtained, and this probably accounted for the confidence with which they associated genes with chromosomes.

Bridges' work with nondisjunction set the pattern for the development of cytogenetics in Drosophila, a few other animal species, and many species of plants. The essential method was to study the association of anomalies and aberrations detected by breeding experiments with cytologically detectable aberrations of chromosomes. When a chromosome from one stock of Drosophila appeared to suppress crossing over among genes previously mapped, it was found to have one sequence of genes in a reversal or inversion of the normal order, such as a-fedcb-gh (cf. Sturtevant and Beadle, 1936). The assumption was made that if the guiding principle in synapsis, when crossing over should occur, is pairing affinity between homologous genes, a with A etc., then genes in such inverted sections can only find their partner genes by formation of loops. Crossing over within such loops will result in a certain number of broken chromosomes so that crossover gametes lead to inviable combinations. Later, the predicted chromosome configurations were found to be associated with genetically identified inversions.

Apparent absence of a normal allele in one parent so that a recessive gene from the other parent was expressed in the hybrid was shown in several cases to be due to deficiency or absence of the segment of chromosome in which the allele had been mapped. In one case analyzed by Bridges (1917), one whole group of linked genes showed this form of pseudo-dominance, and this was shown to be due to the complete absence of one of the small fourth chromosomes. The group of genes which showed the anomalous behavior was thus proved to be located in the fourth chromosome. Similarly, certain genes which normally showed linkage with one group, were shown in anomalous cases to have acquired a new linkage association with genes of a different group. It was subsequently proved by Bridges (1923) that the cause lay in a physical translocation of a piece of chromosome II to chromosome III.

Eventually it became possible to associate each of the four groups of linked genes with one of the four pairs of chromosomes, and, in some cases, to narrow the location of particular genes to small segments of chromosome.

A case studied by Bridges (1921) provided the clinching evidence that one of the groups of linked genes was physically located in the small fourth chromosome. Abnormal flies, with slender bristles, of the type known as Diminished (now called Minute) were mated with flies otherwise normal but homozygous for certain recessive genes which had been shown to be linked with each other but not with the first sex-linked group or with either of the second or third linkage groups. Among the offspring of this cross, those with the Diminished character showed also the recessive alleles of the fourth linkage group. This pseudodominance of recessive genes showed that no normal alleles of these were transmitted by the Diminished parent, and cytological examination of such flies showed that their cells contained only *one* of the small fourth chromosomes. They were called Haplo-IV. This proof that the characteristics of Diminished are due to absence of one fourth chromosome showed that the assumption of a physical deficiency made by Bridges to explain localized aberrations in other chromosomes was well founded. By this method all genes of the fourth linkage group were shown to be physically located in the fourth chromosome. This was further confirmed by combined genetical and cytological study of another aberration, Triplo-IV, in which certain external characters, inherited as a syndrome, were associated with the presence of three fourth chromosomes. When a Triplo-IV fly carried a recessive allele of the fourth linkage group, this allele was transmitted to offspring in the abnormal ratio expected if the allele was borne in the fourth chromosome.

Both types of aberration had arisen by nondisjunction of the fourth chromosome, which contains so few genes that loss of one set, although producing abnormal external characters, still permits the survival of the individual. The change in phenotype caused by the loss or addition of a whole chromosome became the point of origin of a theory (genic balance) which attributes the development of each character to the united action of many genes, some tending to increase and others to diminish the expression of that character. The balance of these opposed tendencies, which is characteristic of a normal or wild type, is upset when the ratio among the genes is changed by loss or addition of chromosomes or segments. This will be discussed in the chapter on the "physiology" of the gene.

Drosophila and the Morgan School

The extraordinarily rapid expansion of knowledge of the physical basis of heredity in the decade 1910–1919 was due in large measure to the thorough and detailed study of one species—*Drosophila melanogaster*. As was mentioned earlier, it was initiated by one group of collaborators working intimately together with simple, even primitive, facilities, in an atmosphere of informality and easy collaboration which became characteristic of much of American research in genetics.

Much of this was due to Morgan, the intellectual and spiritual leader of the group who encompassed all that transpired with the depth and breadth of his insight and his critical faculty. In addition to his gifts as a scientist, Morgan had great literary skill. He described the results and the interpretations reached by agreement among the members of the team with ease and fluency. The dozen books on genetics of which he was sole or joint author represent less than half of the great volume of writing which he achieved, for many of the books had to be revised and reissued in new editions. And he seems to have been continuously engaged in this and in frequent research reports and lectures from many platforms.

His writing was uneven, and many of his papers give the impression of being first drafts dashed off in a hurry and not revised. The enthusiasm with which he reported a new discovery made in his laboratory or attacked misconceptions entertained by his colleagues elsewhere always lent an air of vitality and immediacy to what he wrote.

He depended upon and enjoyed the association with Bridges and Sturtevant and with later students and collaborators. In his memoir of Bridges and other accounts (Morgan, 1941), he revealed the influence which Bridges exerted not only by his incisive proofs of essential points (nondisjunction, deficiency, translocation, sex-balance), but also by his organization of and devotion to the development and maintenance of the living material upon which the whole development of the theory of the gene depended. Morgan tells us that Bridges spent much of his time judging and locating the mutations which continually occurred and combining them into special stocks for special purposes. At the time of Bridges' death in 1938, some 900 such stocks were being maintained. A systematic reporting of these and of research progress with Drosophila generally had been undertaken by Bridges with the collaboration of M. Demerec in the cooperatively maintained and distributed Drosophila Information Service (begun 1934).

This should remind us how important the role played by Drosophila in the development of modern genetics was. Its peculiarity lay in being so remarkably well adapted to the purposes of the human beings who exploited it. As their purposes changed, the animal proved capable of giving different evidence to answer the new questions put to it. Using it, they were able to elucidate the mechanism of Mendelian heredity and provide detailed correlation of genetical and cytological structure and behavior even to the level of fine structure and the analysis of the mutation process. It also provided material for some essential steps in the formulation of ideas in physiological and developmental genetics, and by way of the chemical and physical study of its giant salivary chromosomes it led the way toward the transposition of genetics to a molecular level. It proved to be good material for population genetics and phylogenetic studies; and it also provided for analysis instances of non-chromosomal inheritance on the boundary line between infection and heredity. It has held the center of the stage in genetics for over fifty years, its bibliographic record now having passed 10,000 titles, the number added in the last decade having increased over those in the previous ones.

I conclude this section with a review of the work and contributions of Morgan, the chief architect of the theory of the gene. Morgan's address as president of the Sixth International Congress of Genetics (Ithaca, New York, 1932) was entitled "The Rise of Genetics." It reveals much concerning his views of the origins and aims of genetics and his philosophical attitude toward scientific work. His attitude, of course, influenced his own work in genetics and thereby that of a decisive period in its history. He hated mysticism and obscurantism; this underlay his anti-clericalism. He intended always to keep "within the bounds of mechanistic interpretation." His hope was to focus as directly as possible on efficient tests of hypotheses and to put aside questions about which he could not frame such hypotheses. He gave his admiration and confidence only to what he called "radically experimental" proofs.

Mendel was, then, a man after his own heart, for he had framed simple, direct questions and answered them with quantitative experimental data. Morgan said of Mendel:

> It is the orderly result of disjunction or segregation that is the important feature of Mendel's work; and finally, the clearness with which Mendel stated and proved the interrelations between character-pairs in inheritance, when more than one pair is involved, places his work distinctly above everything that had gone before.

And then, remembering that Mendel was a priest and therefore subject to suspicion of mysticism, he added, certainly with a twinkle in his eye:

> Nevertheless the genial abbot's work was not entirely heavenborn, but had a background of one hundred years of substantial progress that made it possible for his genius to develop to its full measure.

Morgan attributed the beginning of genetics to "the resurrection of Mendel's paper in 1900," but immediately pointed out some of the events in the 100 years of preparation referred to above.

He saw as the most direct ancestry of genetics the work of the plant hybridizers of the nineteenth century which had been made possible by the experimental proof of sexuality in plants by Camerarius in 1694. It was characteristic of Morgan to choose for citation only those bybridizers who worked with Mendel's material, the pea plant, and found new experimental facts which resembled those established and interpreted by Mendel: Knight, Goss, Seton, and Laxton in England, and Naudin in France. The development of what Morgan called "cellular morphology" was important, too, but the history of cytology which he thought necessary to mention is condensed to three sentences and ten names. Weismann's theoretical contribution on the other hand warrants a whole paragraph, its essential effect being to turn attention to the maturation divisions, which formed an indispensable link between Mendelism and the theory of the gene developed by Morgan. The only mention of de Vries is in connection with the mutation theory.

This paper reminds us also that Morgan was a geneticist *pro tempore*. He came from experimental embryological studies in 1910 and returned to them in 1928, never, of course, having left them completely, for in spite of all handicaps to investigation, the causes of change in the developing individual were what fascinated him:

> despite its highsounding name of *Entwicklungsmechanik* nothing that was really quantitative or mechanistic was forthcoming. Instead, philosophical platitudes were invoked rather than experimentally determined factors. Then, too, experimental embryology ran for a while after false gods that landed it finally in a maze of metaphysical subtleties. Perhaps my disappointment at the outcome of the work has led me to an overstatement of its failures.

When it came to looking for lessons from genetics he had this to say: "The theory of balance between the intracellular products of the genes is the most direct contribution genetics has made to physiology"; and with respect to evolution: "Genetics has made a very important contribution to evolution, especially when it is recalled that it has brought to the subject an exact scientific method of procedure." The most important problem for the future seemed to him to be "the physical and physiological processes involved in the growth of genes and their duplication. . . . The ability of new genes to retain the property of duplication is the background of all genetic theory."

This was Morgan's valedictory in genetics, for although one further book, *Embryology and Genetics*, appeared in 1934, Morgan's own experimental work in this field had ended several years before. The last paragraph in his paper of 1932 purported to contain Morgan's advice to his colleagues and successors. It is, in fact, a statement of Morgan's own scientific creed:

> Should you ask me how these discoveries [*i.e.*, those of the future] are to be made, I should become vague and resort to generalities. I would then say: By industry, trusting to luck for new openings. By the intelligent use of working hypotheses (by intelligence I mean a readiness to reject any such hypotheses unless critical evidence be found for their support). By a search for favorable material, which is often more important than plodding along the well-trodden path, hoping that something a little different may be found. And lastly, by not holding genetics congresses too often.

– Chapter 16 –

CYTOGENETICAL ANALYSIS

COMBINED CYTOLOGICAL and genetical analysis of other species, especially certain plants, developed rapidly after 1915 parallel with the Drosophila work. A. F. Blakeslee and his group at Cold Spring Harbor, New York, found in the Jimson weed, *Datura stramonium*, twelve distinct mutant types each containing an extra chromosome in one of the twelve sets characteristic of the normal diploid. One of these trisomic (2n+1) mutants, Poinsettia type, transmitted a gene mutation (purple) in trisomic ratios rather than in the diploid ratios of 3:1 from heterozygotes. This was taken as proof that the gene for purple was located in the chromosome which was trisomic in Poinsettia. Other types of chromosomal aberrations— 2n—1, triploids, tetraploids, and translocations—were quickly recognized and studied by Blakeslee and John Belling (reviewed in Blakeslee, 1941).

An hypothesis put forward by Belling (1926) to explain certain cytological configurations in Datura proved to be especially fruitful. This was called segmental interchange. The assumption was that two equal-armed nonhomologous chromosomes might as a rare aberration each undergo breakage near the median spindle attachment and then mutually exchange the detached acentric segments. The idea was, of course, the same as reciprocal translocation. Since no genes were lost in the process, the rearranged chromosomes might thus become characteristic of certain races or species.

Segmental Interchange

In Datura, for example, a chromosome with arms 1.2 (the dot represents the centromere) may exchange with a non-homologue, say 17.18, and these two chromosomes in a descendant race might then be 1.18 and 2.17. An essential feature of Belling's hypothesis is that what causes homologues to pair at meiosis is attraction (pairing affinity) between like sequences of genes. This would bring together at synapsis like chromosome ends, such as 2 with 2. This led to the prediction that a hybrid between a race with 1.2 and 17.18 chromosome ends and one in which the arrangements are 1.18 and 2.7 should show a ring of four chromosomes at meiosis as follows:

$$1.2 \leftrightarrow 2.17$$
$$\updownarrow \qquad \updownarrow$$
$$1.18 \leftrightarrow 18.17$$

(chromosome-end pairing is shown by double arrows). The prediction was validated by many different tests in Datura. In Drosophila, where gene-markers were more numerous than in Datura, it was validated both by breeding and by cytological evidence.

The hypothesis of segmental interchange proved to be one of the essential steps in resolving the puzzle of the Oenothera "mutants" discovered by de Vries and his successors. Of the other steps, one was supplied by Otto Renner's genetical analysis beginning in 1917, which showed that most Oenotheras are, in fact, complex hybrids. Such a plant breeds true because it produces only two types of gametes, each containing a different complex of closely linked genes which are lethal when homozygous. The genes in all seven pairs of chromosomes thus behave as if they belonged to a single linkage group.

A second step was Muller's proof (1918) of the operation of a balanced lethal system in Drosophila in which a race with two lethals bred true (or nearly so) because the non-allelic lethal genes were on homologous chromosomes, and crossing over between them was suppressed by an inversion. Homozygotes for each lethal gene failed to survive and such stocks were maintained by the surviving heterozygotes. Muller pointed out that permanent hybrids like *Oenothera Lamarckiana* could be maintained as balanced lethal systems if crossing over between homologues was reduced almost to zero. Exceptions due to rare crossovers would then, because of their rarity, be interpreted as mutations.

The third step was Ralph E. Cleland's demonstration (1922, 1929) that in most species of Oenothera the seven pairs form one large ring at meiosis. When the reducing division occurs, alternate members of the ring (which have come from opposite parents) go to one pole and the balance to the other, so that no recombination of genes or chromosomes ordinarily occurs and the whole complex of each parent with its lethals is transmitted as a unit.

The working out of the complex Oenothera story by Renner, Cleland, Sturtevant, S. Emerson, and others was one of the accomplishments of the 1915–1938 period. It has recently been reviewed by Cleland (1962). Actual genetical evidence that segmental interchange in fact involves the exchange of segments between nonhomologous chromosomes was supplied by Drosophila (Dobzhansky, 1929a, 1929b) for which well-marked, mapped chromosomes were available. Applications of this to questions of mutation and evolution will be outlined in other sections.

The power of cytogenetical analysis was greatly increased by two developments which occurred within a few years of each other. One was the discovery, by Muller in Drosophila (1927) and by Lewis Stadler in maize and barley (1928), that ionizing radiations increase enormously the frequency of mutations and of chromosomal aberrations (*cf.* p. 168). This discovery made it possible to produce large numbers of variants for genetical and cytological analysis.

The other development was the intensive study of the giant chromosomes which were found to be regular features of the nuclei of the salivary glands of Drosophila and other dipteran larvae. These chromosomes were first used by Theophilus Painter (1933) to provide a detailed cytological map of the X chromosome of *Drosophila melanogaster*. He showed that the succession of deeply staining bands or discs charged with deoxyribonucleic acid corresponds to the linear order of blocks of genes in the linkage map of this chromosome. The correlation was proved by identifying, in the salivary chromosomes, translocated or inverted sections containing gene loci the relations of which had been worked out in breeding experiments. This was the beginning of an unexpectedly detailed verification of most of the assumptions of the chromosome theory, carried out largely by C. B. Bridges, who provided "salivary maps" for the chromosomes of *Drosophila melanogaster*. The major assumptions derived from breeding analysis had of course already been validated in the cases of duplication, Haplo-IV, deficiency, and translocation by the study of ordinary somatic chromosomes and the construction of rather rough cytological maps showing locations of some of the genes. This work of Painter and Muller (1929) and Dobzhansky (1929a, 1929b) had furnished the first essential validation of the hypothesis of the linear order of loci.

Now, however, the refined analysis made possible by the ability to identify several thousand bands resulted in narrowing the location of a gene to within one or a few bands (*e.g.*, Slizynska, 1938). These were spectacular achievements, but in general chiefly confirmations of the ideas about the structure of the genetical system which had been gained from breeding analysis.

In two cases, however, the study of the salivary chromosomes contributed to more than technical refinements of the theory of the gene. From a detailed study of the band patterns in *Drosophila melanogaster*, Bridges (1935) showed that certain sequences recur in different regions and that these tend to be found paired in the form of synapsis of somatic chromosomes which regularly occurs in Diptera. He referred to these regions as "repeats" and assumed that they had arisen as duplications of the same segment. In this process, the duplication had become inserted near the site of origin (tandem repeats) or at other points in which breaks in the chromosome had occurred and had become incorporated as a transposed piece when the break healed. It was in this way, Bridges thought, that the number of genes in a species might increase.

This idea had been expressed originally by Bridges (1918) upon his first discovery of duplications, the main interest of which (Bridges, 1935, p. 64) "lay in their offering a method for evolutionary increase in lengths of chromosomes with identical genes which could subsequently mutate separately and diversify their effects." Instances of areas deleted and inserted elsewhere were given a similar interpretation by Muller. This view of a mechanism of evolution of the genotype by duplication within chromosomes tended subsequently to supersede the earlier hypothesis that increase in genic content occurred by duplication of a whole chromosome in one set, as in Triplo-IV, or by duplication of sets as in polyploidy. Internal duplications, especially those involving few elements, would run less risk of upsetting the balance among all the genes and thus of being eliminated by natural selection.

The second important theoretical contribution from the salivary chromosome analysis was the proof of the cytological basis of a case of position-effect. This concept had originated with Sturtevant and Morgan (1923), who showed that changes in the Bar eye mutation in Drosophila, such as reversion to wild-type eye form or to a more extreme form of Bar eye (ultra-Bar), are accompanied by crossing over near the map locus of Bar. Sturtevant's 1925 analysis suggested that this crossing over occurred in such a way that it could result in two Bar genes (in effect a duplication of that locus) on one chromosome and no Bar gene on the homologous chromosome. The first showed a more extreme effect of Bar, as in the ultra-Bar $\frac{BB}{+}$ than when the two Bar genes were on opposite chromosomes, as in Bar females $\frac{B.}{B}$ This was probably due to the association of two Bar genes next to each other, an effect of their position relative to the neighboring genes. Bridges (1936) showed that in the region of Bar in the salivary gland X chromosome, one set of bands was represented once in female wild-type flies "reverted" from Bar, twice in Bar females, and three times in ultra-Bar. The Bar eye phenotype is thus due not to a gene mutation but to a duplication (triplication) of the same gene materials, which have different effects when their position is altered.

Although at that time no other case of position effect was proved with the same degree of conclusiveness as the Bar case, partly because of the precision of the salivary chromosome analysis, still it called into question some of the fundamental concepts of genetics which rested on the chromosome theory. One of these was the very discontinuity of the genetic system of the chromosomes, consisting as they evidently do of gene loci sufficiently discrete to be separated and recombined by crossing over or by other forms of chromosome breakage and reconstruction. As Dobzhansky pointed out (1936, p. 382):

> And yet, it is a continuum of a higher order, since the independence of the units is incomplete—they are changed if their position in the system is altered. A chromosome is not merely a mechanical aggregation of genes, but a unit of a higher order . . . the properties of a chromosome are determined by those of the genes that are its structural units, and yet a chromosome is a harmonious system which reflects the history of the organism and is itself a determining factor of this history.

This revelation about the presence of a synthetic or integrative level above that of the gene came at a crucial time and did in fact mark a turning point in genetics, although not in the manner supposed by such opponents of the gene theory as Goldschmidt. The chief efforts in the construction of that theory had been directed toward analysis and had succeeded beyond any expectation that could have been entertained in 1910. But now the tide was to turn and the functional aspects of the components into which the genetical system could be resolved were to receive more and more attention. The structure and functioning of chromosomes, the activities of genes in controlling metabolism and development, and the integration of genes into the genotypes of populations such as races and species could now become actual problems of experimental study.

The Cytogenetics of Maize

The provision of cytological proofs of basic postulates of genetics was accomplished most fully and convincingly in plants by studies of the corn plant *Zea mays*, which by 1915 had become one of the chief materials for genetical research. The contributions of a cooperating group of young investigators led by R. A. Emerson at Cornell University were summarized in two notable publications in 1935: the linkage studies by Emerson, G. W. Beadle, and A. C. Fraser; the cytogenetics of maize by M. M. Rhoades and Barbara McClintock. The latter publication marked the highest point attained up until that time in unifying cytological and genetical methods into a single clearly marked field. Some 400 genes had been identified and most of them placed in ten linkage groups corresponding to the ten haploid chromosomes. The chromosomes of maize as seen in meiosis and especially in the pachytene stage proved to be remarkably favorable for cytological study, and, in 1929, McClintock had shown that all ten chromosomes are morphologically distinguishable. It soon became possible to identify each linkage group with one of the ten chromosomes and to assign certain genes or sequences to definite positions within the physical chromosome.

This was made possible by the identification of plants trisomic (2n + 1) for a specific chromosome. When the genes in a specific linkage group segregated in trisomic ratios in such plants, that linkage group could be assigned to the specific chromosome which was trisomic.

Association of specific genes with specific structures in a chromosome was facilitated by the discovery of visible deficiencies, for example, at the end of a chromosome, and corresponding evidence of absence of a gene or genes. Deficiencies, translocations, and inversions were all recognized and the genetical and cytological findings correlated.

It was at this time, in the early 1930s, that the first cytological maps of Drosophila chromosomes began to appear. The parallelism between the two species was emphasized when cytological proof of genetical crossing over was provided, for the first time in any species, by Harriet Creighton and Barbara McClintock (1931) for maize and by Curt Stern (1931) for Drosophila. Cytological and genetical proof of crossing over between chromatids (in the four-strand stage of meiosis) was provided by Creighton (1932) and McClintock (1933) to supplement the inferences based on genetical crossing over between attached X chromosomes at a four-strand stage in Drosophila.

Moreover, cytological discoveries in maize provided evidence that the meiotic mechanism itself was controlled by the genotype. Thus Beadle analyzed such abnormalities as failure of synapsis, or failure of chromatids to disjoin, as due to single mutant genes.

The cytological demonstrations from maize completed the detailed proof of the chromosome theory and laid the groundwork for later studies of the relations between the structure of the genetic material in the chromosomes and its function in development.

Genetical Cytochemistry

The intense cytogenetical activity of the 1930s awakened the interest of many geneticists in the chemical constitution of chromosomes and genes; and collaborative work between biochemists and geneticists began in earnest in that period. R. Feulgen and H. Rossenbeck in 1924 had discovered a specific color reaction for deoxyribonucleic acid (DNA) and by 1937 had shown that this substance was largely confined to the chromosomes, while another pentose nucleic acid (RNA) was in the cytoplasm. Chromatin from that time could have been called DNA and chromosomes "DNAsomes."

Using absorption spectra to determine the chemical constitution of the bands of the salivary chromosomes which cytologists had found to stain deeply, Caspersson (1936) showed them to contain excess pentose nucleic acid. Subsequently, the material was identified as DNA by chemical methods. The probability became strong that the genes, known to be in these bands, were associated with DNA. At the same time it was shown that crystalline viruses, endowed like genes with the power of self-replication, also are nucleoproteins. It appeared likely also that the wavelengths of ultraviolet light most potent in producing mutations were those absorbed by DNA.

The correlation between growth of the salivary chromosomes and increase in nucleic-acid content, and the changes in the quantity of DNA when, in chromosome rearrangements, euchromatic regions are brought near to heterochromatic ones (Schultz and Caspersson, 1939), suggested that position effects known to occur in the latter cases are related to local changes in the nucleic-acid metabolism. This work of the late 1930s became the basis for correlations established later between the genetic effect and the physical and chemical constitution of a chromosome region.

– Chapter 17 –

THE EXPERIMENTAL STUDY OF MUTATION: 1910–1938

THE CHIEF ADVANCE made in genetics in the 1920–1930 decade was the convincing proof that mutations could be induced en masse by radiations of different kinds. Since the mutants so produced were of the same kinds which had been found previously and had supplied the materials for Mendelian analysis and the mapping of chromosomes, it was evident that increases in the amount of energy absorbed by the hereditary material resulted in increases in the rate at which mutations occurred in untreated animals and plants.

The ideas and terminology underlying this proof had come mainly from de Vries, who had shown by his discoveries of mutant forms of Oenothera that it was possible to study the process experimentally and that there was an order that could be analyzed.[20] But the heritable changes which de Vries had

---

[20] Åke Gustafsson, in an historical sketch (1963) of the induction of mutations, has called attention to the foresightedness of de Vries who, in 1901, had written: "Knowledge of the principles of mutation will certainly at some time in the future enable a fully planned artificial induction of mutations, *i.e.*, the creation of new properties in plants and animals. Moreover, it is likely that man will be able to produce superior varieties of cultivated plants and animals by controlling the origin of mutations." In 1904, in a lecture at the Station for Experi-

observed had been shown to consist of a heterogeneous collection, most of which did not represent new gene forms but rearrangements of existing ones in chromosomal changes (polyploidy, heteroploidy, etc.), some of them given the appearance of rarity by suppression of recombination within systems of balanced lethals. The discovery and study of the first Drosophila mutants, many of which were alleles of others already known, led to recognition of "gene mutation," *i.e.*, the origin of new alleles, as the prime source of new variants in evolution. H. J. Muller had already pointed this out when, in 1918, he began to study the normal spontaneous process of mutation in *Drosophila melanogaster*. This led to his proof, in 1927, of what he called "Artificial Transmutation of the Gene." The major feature of this proof was an increase in mutation rate of some 1,500 times in the progeny of Drosophila males exposed to heavy doses of X rays.

This was not a chance discovery but the result of a deliberate theoretical analysis and a well-prepared novel experimental design adapted to test a specific question. This question was whether mutational changes could be deliberately induced and accurately and quantitatively detected. Previous efforts to do this had not led to clear and decisive results. The reasons for failure had been due, as N. W. Timoféeff-Ressovsky (1935) pointed out, to the failure to meet five essential conditions: genetic purity of the material to be tested, large numbers of individuals in test and control material, genetical methods for detecting newly arisen mutations, exact methods of analysis of different types of variations found, and knowledge of the manner of action of the agent.

In 1921, Muller had begun his talk at an international congress in New York: "Beneath the imposing structure called Heredity there has been a dingy basement called Mutation." He was then devoting his major efforts to providing the five

---

mental Evolution at Cold Spring Harbor, de Vries suggested, according to the report by Blakeslee (1936), that "the rays of Roentgen and Curie, which are able to penetrate into the interior of living cells, be used in an attempt to alter the hereditary particles in the germ cells."

For a fuller historical account of gene mutation, see Stubbe (1938, 1963) and Timoféeff-Ressovsky (1937).

conditions just mentioned. He outlined the problem at another meeting the following year in a penetrating and prophetic paper. Here he expressed his recognition that detection of change in the individual gene was a method of measuring normal mutation frequency on a very large scale. The method devised was to have a sure means of identifying a given chromosome by a marker gene, to prevent it from acquiring a new gene by crossing over from a homologous chromosome, and then to bring it into such combination as would reveal any mutant change in it which had arisen either spontaneously or after treatment.

The most efficient method was one designed to detect new lethal mutations in the X chromosome. Preliminary experiments in which cultures were raised at two different temperatures served to test the method and gave estimates of two to five new lethals per 10,000 X chromosomes tested. In later experiments in which males were treated with X rays, the mutation rate in the X chromosome was increased many fold. Thereafter, there remained no doubt that the mutation rate could be raised intentionally in a manner generally proportional to the intensity of the radiation applied. Mutations recovered were not only changes apparently localized in individual genes but also a variety of chromosomal aberrations. This was quickly confirmed in all essentials for other animals and plants. Radium-induced mutations in Datura had, in fact, already been found by Blakeslee and C. Stuart Gager. Independently of Muller's work, Stadler showed that short-wave radiations induced gene mutations and chromosome abnormalities in the progeny of treated plants and seeds of barley and maize.

The intense activity in the study of mutation touched off chiefly by Muller's demonstration brought into being a new branch of genetics. After 1930, radiation genetics expanded rapidly and came to occupy a central position, not merely by providing new variants for analysis, but in its own right as a mode of analyzing the basic structure of the hereditary material in the chromosomes. An important factor in this was the fact that mutations induced by a variety of agents—X rays, gamma rays, beta rays, cathode rays, and ultraviolet light—belonged to the same categories as the mutations which had appeared spontaneously. This provided assurance that it was in fact the mutation process responsible for evolution which was being studied. The study of spontaneous mutation was revitalized, and Stadler's demonstration of differences in the spontaneous mutability of different genes in maize, following upon Demerec's demonstration of highly mutable genes in *Drosophila virilis,* stimulated the search for causes of such control. Rhoades found a very potent controlling gene which caused high mutability of one locus but just to one particular allele (from a_1 to A_1). Since both the stimulating and the target allele were inherited independently, it appeared that the difference between stable and unstable genes was not a fundamental one, but that these were extremes of a continuous spectrum. It also suggested that since action at a distance within the same nucleus affected mutability, then the one-hit-one-mutation theory, the so-called "Treffer theory" devised to account for the proportionality between radiation dose and mutation rate, could not be the only correct interpretation. Further development of this idea of genic control of mutability appeared later in the proof by McClintock of genic activators and controller systems within the cell. Doubts about the adequacy of the Treffer theory were increased when it was shown that mutation rates (in bacteria) could be increased by irradiating not the cells but the medium in which untreated cells are grown.

At the end of the period the "dingy basement" had been subjected to increasing illumination. Many facts had been revealed, perhaps none more important than the complexity of the process and the wide varieties of conditions affecting it.

Changes in Genetical Theory: The Study of Gene Mutation

The corpuscular view of the gene was justified by the results obtained from its application. This view required emancipation from the restrictions imposed by Bateson, Johannsen, and those who feared the restoration of the morphological concept of living units a la Darwin, Naegeli, and Weismann. Muller's conception of the gene as a material particle had, in fact, a liberating effect. If the gene were real, one could hit it with a projectile such as a free electron and get an estimate of its size. The concept of the corpuscular gene, which took form largely from the Drosophila work and reached its climactic stage in the "hit theory" of K. G. Zimmer, was a very different structure from the gemmules or ids as conceived of in the pre-Mendelian period. The new concept of the gene was derived by the operative reasoning stated by Morgan in the *Theory of the Gene* and validated by cytogenetic methods. But its chief prophet was Muller, who declared his independence of the Morgan group with which he had begun and increasingly blazed his own trail and built his own school. In 1921, he said:

> It is not mere guesswork to say that the genes are ultra-microscopic bodies. For the work on Drosophila has not only proved that the genes are in the chromosomes, in definite positions, but it has shown that there must be hundreds of such genes within each of the larger chromosomes, although the length of these chromosomes is not over a few microns. If, then, we divide the size of the chromosome by the minimum number of its genes, we find that the latter are particles too small to give a visible image.

In the same paper he had this to say about the problem of gene mutability:

> The most distinctive character of each of these ultra-microscopic particles—that characteristic whereby we identify it as a gene—is its property of self-propagation: the fact that, within the complicated environment of the cell protoplasm, it reacts in such a way as to convert some of the common surrounding material into an end-product identical in kind with the original gene itself. This action fulfills the chemist's definition of autocatalysis; it is what the physiologist would call growth; and

when it passes through more than one generation it becomes heredity. It may be observed that this reaction is in each instance a rather highly localized one, since the new material is laid down by the side of the original gene.

He called attention to the similarities of genes and bacterial viruses, the then newly discovered "d'Herelle bodies":

It would be very rash to call these bodies genes, and yet at present we must confess that there is no distinction known between the genes and them. Hence we cannot categorically deny that perhaps we may be able to grind genes in a mortar and cook them in a beaker after all. Must we geneticists become bacteriologists, physiologists, chemists, and physicists, simultaneously with being zoologists and botanists? Let us hope so.

Imagining the gene as a material body was, in Muller's case at least, a creative act which resulted in a great expansion of the horizons of genetics. Far from narrowing the focus to a "bead on a string," as some biologists feared it would, it had just the opposite effect and brought the powerful force of the rigorous and imaginative thinking of the new physics into genetics. This influence was felt in many ways, chiefly in directing attention to substances which, by their response to mutation induction, might reveal the chemical and physical composition of the gene. By 1940, the substances pointed to were nucleic acids, particularly DNA, which absorbs light of the wave-length most effective in producing mutation.

Mutations in Human Genes

While there has never been much doubt that differences in human genes originate by mutation as in other organisms, evidence that this is so can be obtained only by indirect arguments which by themselves would not be convincing. The prototype of such arguments was given in 1921 by C. E. Danforth (1923), when he proposed to estimate mutation rates of dominants such as polydactyly and syndactyly by assuming their equilibrium frequency to be determined by the product of the mutation rate and the average number of generations that the mutant persists in the population. Later Haldane (1935) and Gunther and Penrose (1935) estimated mutation rates of hemophilia and epiloia. However, there is no independent check on their results nor, in the case of hemophilia at least, any assurance that the recurrences could be referred to the same gene.

The rapid progress of genetics in the 1920s and 1930s had important philosophical consequences which tended to spread from the special field and to affect biologists generally. One of the chief of these was the recognition of the enormous genetic diversity of living systems; diversity not only between species but within the species and even in the local population. At the beginning of genetics, genetic variations were curiosities, treasured as useful for studying inheritance, but at the end of forty years of genetical analysis, the numbers of genetic elements recognizable in any species had become so huge that identity, or lack of difference, between two individuals of a random-breeding population had become recognized as the unexpected, in fact the non-existent, condition. In the early part of the period, domesticated plants and animals provided the materials for analysis of the variety which was at that time obvious. Estimates of the amount of diversity continually expanded as large and more inbred populations were subjected to close observation in the laboratory, as in the case of Drosophila. Ability to recognize genes with lethal effects, genes with small effects, multiple alleles, and polygenes not individually identified, resulted in estimates of hundreds or thousands of variable elements, even before it was found possible to produce new mutations in a wholesale fashion by radiations and other mutagens. The Drosophila species, in their actual and potential variety, were proved to have the normal pattern of cross-fertilizing populations. Thinking in terms of "types" had to be replaced by thinking of a multitude of genotypes, a variety held in some balance between opposing influences. Human populations were no exception to this, and recognition of the genetic uniqueness of each human individual began, just before the Second World War, to call for reassessment of many traditional ideas about man and his place in nature.

Also, there was exhibited in genetics a sign of scientific maturity in the deliberateness of planning of scientific investigation. Muller's experimental design for large-scale mutation studies had a notable influence in this direction. The results demonstrated that confidence in the knowledge of the genetic system acquired by experiment was well justified, and the experiments demonstrated a mastery of technique which was to render many old problems susceptible to study and to new solutions.

– Chapter 18 –

THE "PHYSIOLOGY" OF THE GENE

As GENETICS came to maturity in the 1930s, the logical system and the technical methods by which the genetical
system in the chromosomes was described tended to over-
shadow another aspect of heredity which had appealed to
perceptive biologists at the end of the nineteenth century.
This was stated by E. B. Wilson in his book of 1896, *The Cell
in Development and Heredity*. "Inheritance," Wilson wrote,
"is the recurrence in successive generations, of like forms of
metabolism." Its essence is repetition of biological activities of
very specific kinds. The characteristics of individuals and
species are maintained by continuous acts of repetition of patterns at each cell division. This implies the control by the
genotype of repeated syntheses of the same variety of chemical
structures in each organism and cell. The genotype must likewise control the norms of reaction by which each individual
and species attains its characteristic bodily form. Such questions again claimed attention in the two decades ending in
1938.

Only two concepts which emerged from the work and
thought of that period will be chosen for discussion out of
the large number reviewed in Goldschmidt's *Physiological
Genetics* (1938) and Wright's *Physiology of the Gene* (1941).

These are the concept of genic balance, as formulated in the early twenties, and a theory of the manner in which genes control sequences of metabolic reactions, developed in the late thirties. The former was derived from noting how development was affected by aberrations either involving groups of genes lost from a chromosome section, or involving the loss or addition of whole chromosomes and changes in the ratio among different chromosomes. The development of a theory of genetic control of metabolism depended upon the identification of individual mutant genes in controlled combinations and the measurement of the effects of these on cell constituents or metabolic processes. The present view that genes exert their effects by controlling enzymes derives in large part from earlier work with plant and animal pigments and especially from the analysis of the action of genes controlling eye colors in the flour moth Ephestia and in Drosophila. The history of the ideas connecting genes and enzymes will be reviewed first.

Although the idea that genes control enzymes sounds modern (and is so in the specific form—enzyme synthesis—in which it is held today), it began to be expressed almost as soon as phenotypic differences were traced to individual genes. As noted earlier, Garrod assumed in 1902 that the mutant gene responsible for alkaptonuria had been changed in such a way that an enzyme which normally splits the benzene ring in alkapton had dropped out or become inactive. Cuénot suggested in his first paper on coat-color differences in the mouse (1903) that the effects of genes on pigment might be exerted by control of enzymes, and there were frequently suggestions of this sort throughout the intervening years. Wright, in 1917, gave a speculative interpretation of coat-color variation in mammals in terms of two enzymes controlling respectively black (melanin) and yellow pigments.

Breeding analysis revealed that many interacting genes were responsible for the wide variety of colors of flowers and leaves, and when these were studied chemically by H. Onslow, Muriel Wheldale, Rose Scott-Moncrieff and others, some genes were found to control single steps in the transformations of pigments such as reduction or oxidation (Lawrence and Price, 1940) which would be catalyzed by enzymes.

Genic Control of Steps in Metabolic Sequences

An assumption that a diffusible gene product was responsible for eye pigmentation in Drosophila was made by Sturtevant (1920), after noticing that in gynandromorphs of Drosophila the male parts showed the effects of several sex-linked genes in the single X chromosome inherited from their father, but the recessive eye color, vermilion, which was known to be present in that X chromosome was not expressed in these parts. Sturtevant concluded that the normal allele of vermilion, present in the female tissues of the gynandromorph which had two X chromosomes, had determined the eye color of cells with a single X chromosome in which that allele was absent. A diffusible substance could, of course, be suspected to be responsible. This substance was subsequently found and identified, just before 1940, by the collaborative work of Beadle, Boris Ephrussi, and E. L. Tatum.

At the beginning of a new kind of analysis of pigment formation in insects, Ernst Caspari (1933) showed that a pigment difference in the flour moth *Ephestia kühniella* is controlled by a pair of allelic genes, *A* and *a*, which affect the production of a diffusible substance. He referred to this as a hormone. If a larval testis from an animal with *A* (dark pigment) is transplanted to a colorless *aa* larva, the host develops the color of skin and eye of the implant. The formation of the *A* hormone appeared to be blocked in the *aa* animals.

The study of gene-controlled diffusible substances in insects was opened at Göttingen by Alfred Kühn and his students, of whom Caspari was one. It developed independently toward the same goal as that achieved by Beadle, Ephrussi, and Tatum working with Drosophila. These two groups set the basic pattern for the investigation of the role of genes in metabolic transformations. This work underlay the development of the "one gene, one enzyme" hypothesis, which proved to be a useful guide in working out gene-controlled syntheses in microorganisms.

The design of the Drosophila experiments depended on the prior identification of a large number of mutant alleles affecting the presence and degree of expression of pigments in the compound eyes. The larval buds from which the eyes develop can be transplanted, before they have synthesized their pigment, from one larva to another larva (the "host") in which the implant completes its development.

Eye buds from most of the eye-color mutants develop their own (inherited) eye color when transplanted to hosts containing the dominant allele of the mutant. These genes act autonomously in the eye tissues. There were two exceptions to this rule. Eye buds from the mutant vermilion (v) which lacks the brown pigment of the normal red eye became normal red in a wild-type host and so did eyes from cinnabar (cn), another mutant resembling vermilion. The wild-type host supplied something that was missing in the mutant eyes. Cinnabar larvae were able to supply this substance to vermilion eye buds (since the latter turned red when developing in cinnabar larvae), but a vermilion host could not repair the defect in cinnabar.

The theory devised to explain these experimental results was that the mutation to vermilion had blocked one of the steps in the production of brown pigment and the mutation to cinnabar had blocked a different and later step in the same synthesis. The general pattern appeared to be:

$$p \rightarrow a \rightarrow b \rightarrow c$$
$$v \quad\ cn$$

In the presence of the normal alleles of v and cn the synthesis goes from p (a substrate later identified as tryptophane) to substance a (kynurenine) to substance b (3-hydroxykynurenine) to end product c (a brown ommochrome pigment). If each step is catalyzed by an enzyme, then mutation to v

causes absence or ineffectiveness of the enzyme responsible for conversion of a to b and thus a block to the synthesis. Mutation to *cn* similarly prevents the step from b to c. The model was a purely formal one devised to explain the transplantation results. It antedated the identification of any of the substances as chemical entities. Its function was logical rather than factual, and in this sense it resembled the models, such as the linear order of genes, constructed to express the results of breeding experiments rather than the purely physical relations of particles. It gave rise to the assumption that since only one out of a large number of eye-color mutants blocked a first catalytic step, then the "total specificity of a particular enzyme might somehow be derived from a single gene" (Beadle, 1959). This hypothesis was the basis of later extensive work on nutritional requirements in Neurospora which led to the elaboration of the one gene, one enzyme hypothesis.

When Beadle and Tatum went to receive the 1958 Nobel Prize for this work, Beadle (1959) pointed out that, although they had not known of it at the time of analyzing the eye-color reaction sequence in Drosophila, it was now clear that "we had rediscovered what Garrod had seen so clearly so many years before." Thus, the first ascription of enzyme control to a gene occurred in connection with the first human condition (alkaptonuria) recognized as due to a Mendelizing gene. The further development of this idea belongs to a period beyond the limits of this history.

It should be pointed out, however, that the progress in this direction in the 1940s and 1950s owed much to the introduction in the early 1930s, of a new organism for research in biochemical genetics. The life cycle of a bread mold, *Neurospora crassa,* an ascomycete, had been worked out in 1927 by the mycologist Bernard O. Dodge. He found that the eight spores in the ascus occur in a fixed order resulting from two meiotic divisions and one mitotic (equational) division. He worked out the inheritance of sex and of the first mutant expressed in the color of the spores. Segregation of these after a cross always occurred among the spores in a 1:1 ratio. The pattern in which these were distributed showed whether segregation occurred at the first or second reduction division. This permitted recovery in a known order of all of the products of meiosis and made this an ideal organism for tetrad analysis.

The genetics of *Neurospora crassa* was worked out by Carl C. Lindegren, who, by 1936, had mapped several groups of linked genes and had shown that crossing over occurs in the first of the two meiotic divisions. It was with this organism that Beadle and Tatum worked out the principles by which genes control reaction sequences. Beadle has given an interesting account of this in his *Genetics and Modern Biology* (1963).

Genic Balance

The concentration of attention on the segregation, recombination, and effects of individual genes was an essential condition of progress in the early development of genetics. It often evoked the criticism that the view of the organism that emerged was that of a mosaic structure rather than of a functioning whole. But the very process of analysis produced a more synthetic concept. As soon as the multiplicity of genes was recognized, and cases of the interactions among them were worked out, it became evident that all must act in some concerted way as part of a system. The idea that the normal phenotype of the species represents an equilibrium produced by the opposed effects of different genes took specific form about 1920. At that time it was discovered, both in Drosophila and Datura, that addition or subtraction of one chromosome in the genome produced changes in phenotype even when no changes in individual genes were found. The absence of one fourth chromosome was found by Bridges (1921) to produce in Drosophila a complex of differences from the normal in body size and color, in bristles, wings, viability and many other characters. The interpretation of the effect of deficiency of a whole chromosome in Haplo-IV Drosophila was based on the assumption that each character of the normal wild-type depends on the concerted action of many genes, some affecting its development in one direction, others in the opposite direction. The point of balance between the plus and minus influences on each character is shown by the wild-type. Genes with opposite tendencies can be assumed to be distributed at random among the chromosomes, so that if a section or an entire chromosome is removed, it is likely that more of one kind (plus or minus) than another will be removed. If this is so then more of one kind than the other will be left acting, and a different complex of characters (*i.e.*, a mutant) will result. This assumption was supported by observations on flies having an extra fourth chromosome (Triplo-IV). The characters affected in these were like those affected in Haplo-IV flies, but the deviations from normal were in opposite directions in the two mutant forms. The normal diploid 2n was intermediate between the 2n−1 (Haplo-IV) and 2n+1 (Triplo-IV) conditions.

Nearly forty years later, this theory served as the basis for the discovery that certain abnormal human syndromes, for example, mongolism, were due to chromosomal aberrations.

In the early 1920s, Blakeslee and Belling and their associates found in the Jimson weed, *Datura stramonium*, that plants which are trisomic (2n+1) for one of the twelve chromosomes differ from the normal diploid and from each of the other eleven trisomic types in an easily recognizable set of phenotypic characters. They supposed this to be due to a departure from the normal balance among the genes, those in eleven sets being double and in one set, present three times. This idea was strikingly confirmed when the same chromosome was added once (2n+1) or twice (2n+2) to the diploid, the latter differing in the same characters to a more extreme degree. When the same chromosome was added to a tetraploid plant the 4n+1 and 4n+2 and 4n+3 types differed from 4n in a similar direction but to a lesser degree than when the same extra chromosomes were added to a diploid. The Datura and Drosophila results coming at the same time corroborated and supplemented each other, Datura providing easily controlled and observable chromosomal changes but few identified gene markers, while in Drosophila all chromosomes were well-marked, but only the small fourth could be present singly or triply without causing lethal effects.

The most spectacular application of the genic balance theory, however, was attained in Bridges' (1922) interpretation of sex types in Drosophila. In descendants of triploid (3n) females, there were found, in addition to normal males and females, individuals intermediate between males and females (intersexes) and others with female or male somatic characters exaggerated. The state of the sexual characters was found to depend on the ratio of X chromosomes to sets of autosomes. A 1:1 ratio such as 2X:2A (diploid), 3X:3A (triploid), or 4X:4A (tetraploid), determined a normal female. A 1X:2A (= .5) ratio gave rise to a normal male; an intermediate ratio 2X:3A (= .67), to a sterile intersex; 3X:2A (ratio 1.5), to a sterile "superfemale"; and 1X:3A (= .33), to a sterile "supermale." The balance theory of sex determination which Bridges derived from his results assumes that sex differences are due to the action during development of two opposed sets of genes, one tending to produce female and other male characters. The two sets are not equally effective, female-tendency genes being more effective in the whole genome since the 2X:2A diploid (or triploid) is female. Male tendency genes are more effective in the autosomes (1X:2A = ♂), while female-tendency genes are more numerous or more effective in the X chromosome. On this theory, the chief factor in determining genic balance is the ratio between X chromosomes and autosomes. The differing gene content within the chromosomes was assumed to have been reached through natural selection acting on hereditary variants (mutations) in such a way as to ensure the fixation of the two stable equilibrium points, i.e., 1 and .5 representing the normal female and male.

The basic ideas involved in the balance theory of sex as stated by Bridges had already been proposed by Goldschmidt, who had been the first to find and interpret intersexes in moths. Later (1927, 1938), Goldschmidt based a general physiological theory of heredity chiefly upon the ideas derived from a study of the development of the sexual types. The gypsy moth intersexes studied by Goldschmidt were produced within the normal diploid genome by combinations of genes from different geographic races. The opposing sets of genes leading in a male or female direction could thence not be identified or associated with specific chromosomes, and the balance between them was ascribed to relative "strengths" which could not be measured directly. Thus, although Goldschmidt was undoubtedly the originator of the balance theory of sex, the convincing nature of the proofs provided by Bridges and others working with Drosophila prevailed in giving that work the greater influence upon the subsequent development of the genic balance theory.

Goldschmidt's extensive and thorough analysis of the development of intersexuality led him to propose a developmental interpretation, which became the prototype of many subsequently employed to explain the action of mutant genes. Goldschmidt assumed that intersexes begin to develop in a female (or male) direction determined first by the sex chromosome constitution ($XY = ♀$; $XX = ♂$) and continuing as such up to a critical period or turning point, after which the genes of the opposite sex prevail and later development is of the opposite sex type.

This idea of a turning point, although never subject to strict proof, was useful in interpreting the fact that in Drosophila triploid intersexes start development as males and end as females. As a gene-controlled timing or switch mechanism, it was incorporated in many generalized theories of phenogenetics in addition to those elaborated by Goldschmidt.

– Chapter 19 –

DEVELOPMENT AND GENETICS

IT SEEMS FAIR to say that no well-supported general theory of the manner in which genes control morphogenetic processes in development had been attained by 1939. Consequently, no detailed history of this field will be attempted. It should be noted, however, that the beginnings of a theory of genic control of enzyme function had provided a model which would underlie applications of gene control in morphogenesis. What was lacking was actual demonstration of first or primary effects of specific genes on elementary developmental processes. That morphogenetic substances emanating from the nucleus were concerned in such processes had been thought probable after Hämmerling's experiments with the unicellular alga Acetabularia. Here grafting experiments between species differing in the form of cap or an umbrella-like part showed that material from the nucleus could pass through foreign cytoplasm to effect its specific action at a distance.

In many cases, embryological descriptions of mutant characters during development were made in birds, mammals, insects, and a variety of plants. Gene mutations served in this work as substitutes for the experimental operations made by morphologists. The questions were generally phrased in embryological or morphogenetic terms and usually concerned development. Geneticists sometimes implied this when they referred to such studies as "Mendelian embryology," genetics supplying the adjective, development the substantive. Embryologists generally did not feel the need of genetics in studying their problems, and indeed the idea that all cells contained the same constant set of genes seemed to many of them to mean that genes as such could not be responsible for differentiation.

Some of the causes of slow progress in developmental genetics may be gained from considering three books published shortly after Morgan's *Theory of the Gene* (1926). One of these marked the end of an old way of looking at problems of heredity and development. The second described a new view based ostensibly on the newly elaborated theory of the gene, while a third attempted a reconciliation between the old and the new. The three books were the fourth edition of Hans Driesch's *Philosophie des Organischen* (1928), Richard Goldschmidt's *Physiologische Theorie der Vererbung* (1927), and W. Schleip's *Entwicklungsmechanik und Vererbung bei Tieren* (1927).

In Driesch's book there appears the following sentence (p. 183): "Thus, without further ado we have our conclusion: the genes are means for the forming of structure, which the entelechy employs."[21] Driesch's retreat into vitalism certainly did not mark a general tendency among experimental embryologists; rather it was an extreme and personal form of a feeling (one might say almost a fear) which was shared by others. To ascribe the direction of development to interactions between material bodies such as genes and material factors of the internal and external environment seemed somehow inadequate and in a sense an undignified way out of a problem of much grander dimensions.

The explanation of development, clearly, must always involve terms representing temporal sequence, integration, individuation, and the like; whereas gene transmission was an either-or, once-for-all affair, and, after 1915, no longer tinged with

[21] "So haben wir ohne Weiteres unser Ergebnis: die Gene sind Mittel für Formbildung, welche die Enteleche benützt."

mystery. The theory of the gene was in fact the quintessence
of mechanism and by its nature was bound ·to appear crass
and naïve and incomplete to a philosophically conscious person like Driesch.

Schleip added to the discomfort of some of the older embryologists when he pointed out what would become of some of
their revered and useful expressions. In genetical terms "*prospektive Potenz*" means "genotype" and "*prospektive Bedeu-
tung*" means "phenotype," while "determination," the other
major word, means the act of transformation of one into the
other—the realization of the "*prospektive Potenz*" in development. Not only did sacred terms get translated into the new
vulgate, but Schleip went on to point out that developmental
physiology had reached a dead end. What he meant by "dead
end," as Barthelmess has noted, was not cessation of good
experimental work, which in fact was increasing, but the fact
that no way could be seen ahead in the interpretation of
regulation and regeneration—the restoration of wholes after
subtraction of parts.

One should not underestimate the effect of the conflict between philosophical and scientific motives which has been a
part of the atmosphere in which embryological work is carried
on. Development, since it results in an individual, always
raises the problem of one out of many, and that can be regarded either as a philosophical or a scientific problem, depending on what one proposes to do about it. Schleip's
recommendation was to use methods of physics and chemistry
which hadn't as yet contributed much to the analysis of
development and especially to identify and study the role of
enzymes in development. This, too, was a recurrent theme in
T. H. Morgan's writings of the 1920s.

There is a certain parallelism between the reaction to genetics on the part of embryologists and the skepticism with which
many biologists whose primary interest was evolution viewed
the first demonstrations of the physical basis of heredity.
Genetics seemed to be dealing with minor, superficial, even
trivial matters which could have little to do with the grand

strategy of evolution. It did not help matters that those who devoted themselves to analytical genetics seemed not to be interested in the larger problems of biology. Goldschmidt spoke for classical biologists generally when he said in the introduction to the *Physiologische Theorie* (1927):

> This fact [the solution of the transmission problem by the Morgan school], however, deludes many geneticists into thinking that here also a theory of heredity has been established, so that they have no interest in investigations which go beyond the working out of the transmission mechanism of heredity.[22]

One part of his meaning was that the Drosophila geneticists had not been much impressed by Goldschmidt's attempt to expand the scope of gene theory to include development.

Goldschmidt's Physiological Theory

Goldschmidt belonged to no school but his own. Although his primary training had been in cytology, he was thoroughly at home in embryology and other classical fields before he became interested, around 1910, in genetics. When he came to develop his *"Physiologische Theorie"* he took the transmission mechanism as it had been portrayed by Morgan and his colleagues and went on to supply what seemed to him to be lacking in that theory, namely, a physiological means for translating genes into characters during development. His theory was thus one dealing with development and not with transmission or structure of the genetic material. Goldschmidt's was not the first attempt in this direction; Valentin Haecker had used the term *"Phänogenetik"* (phenogenetics") in 1912, and the title of his book of 1918, *Entwicklungsgeschichtliche Eigenschaftsanalyse* ("Developmental Character-analysis") ac-

---

[22] Diese Tatsache [the solution of the transmission problem by the Morgan school] aber täuschte vielen Genetikern die Idee vor, dass hier auch eine Vererbungstheorien gefunden sei, so dass ihnen das interesse ihr Arbeiten, die über die Erforschung des Erbsmechanismus hinausgingen, fehlte.

curately described the method, but in fact the text added little of theoretical interest. Haecker, like many others, thought that the goal of such developmental analysis was to learn more about the gene through studying its effects. The expectation was, of course, disappointed. Goldschmidt renounced such a purpose, but in fact his theory implied that attributes of the gene, such as its quantity, could be derived from measuring its phenotypic effects. In many ways a more realistic forecast of research in physiological genetics was Hagedoorn's essay of 1911, "Autocatalytical Substances the Determinants for the Inheritance Characters. A Bio-Chemical Theory of Inheritance and Evolution." In fact, neither of these earlier attempts had much to do with the growth of developmental genetics since there were so few observations by which theories could be tested.

By and large, Goldschmidt's theory suffered the same fate and for the same reason. What kinds of measurements could be obtained to confirm or deny the main thesis that each gene acts through its control of rates of reaction? Examples of mutant genes causing changes in rates are, of course, numerous, but in order to control development as a whole the different reactions have to be attuned; and Goldschmidt's concept of *"Abgestimmte Reaktionsgeschwindigkeiten"* while stating a necessary condition, gave no indication as to what was to be measured or how. Nevertheless, although Goldschmidt's ideas had a considerable influence in directing attention to the developmental processes intervening between gene and character, they did not lead to the establishment of a general theory of development in genetical terms. In fact, at the end of the period, as signalized by Goldschmidt's book of 1938, doubt remained whether there was a field with a defined problem which could be identified as developmental genetics.

It is sometimes forgotten by present-day students that in classical biology—in the latter half of the nineteenth and early twentieth century—heredity included development. A theory of heredity would have been quite inadequate if it gave a view only of the transmission mechanism (such as that provided by

Morgan's theory of the gene) and failed to explain how the repetition of like characters in the offspring was brought about. It was E. B. Wilson, himself the bridge between the older and newer views, who in 1896 defined the essential problem (*cf.* p. 175). The chief theories of the nineteenth century—Darwin's, Naegeli's, Weismann's, de Vries', and others —all dealt with development, although their major concerns were with continuity and evolution. However, embryology and genetics began to diverge after 1900 and theoreticians of development like Wilhelm Roux and Hans Driesch and Jaques Loeb, and, later, Hans Spemann, treated the transmission mechanism as incidental and secondary, if they mentioned it at all.

There persisted throughout the development of classical genetics a tendency on the part of embryologists to consider genes as concerned only with rather trivial characters which appeared after the important events in development had already been determined by the egg cytoplasm. The creators of theory in genetics, on the other hand, generally confined themselves to elucidating the transmission mechanism and the structural features of the genetical material without facing the problem of differentiation. When I was a beginning student of genetics, the controversy aroused by these diverse views was the first in which I participated, and I pointed out (Dunn, 1917) that chromosomal genes could, and in known instances did, control first steps in development, whereas the influence of non-genic materials was usually in passive transmission, rather than in the determination of future events.

When embryonic development constitutes the central problem, our ways of thinking differ rather sharply from those employed when the transmission mechanism of heredity is the subject. Obviously, the former problem has more dimensions than the latter, the most obvious additional dimension being that of time. The biological transmission systems that we know most about consist of particles which have to be dealt with by counting.

The use of numbers exerts on scientists a sort of magic

influence. "Number work" has the property of being right or wrong, with nothing between. When, in the early 1900s, the Mendelian principles attracted converts, most of them came from fields that at that time were primarily descriptive and non-quantitative. Bateson, Castle, Conklin, Cuénot, and Morgan had been trained in embryology, yet when they became interested in genetics their previous interests receded before statistical concepts which were so simple that newcomers and amateurs could use and even extend them.

In their studies, however, the problems about development were displaced by older questions concerned with the mechanism of evolution, for to them genetics was a means of studying the latter. It was evolution, not development, which brought them into genetics. In Morgan's case, genetics was a passing interest which endured for about twenty years, roughly 1908–1928, after which he resumed his work in experimental embryology. But it cannot be said that any of these converted embryologists made major experimental contributions to the problems of developmental genetics.

After his genetical research had ended, Morgan wrote a book called *Embryology and Genetics* (1934). Just as the two subjects remained independent and equal in the title, so they did also in the body of the text; one could detect very little recombination or crossing over between the two sets of problems. In the last paragraph of that book Morgan with his perceptive insight pointed out the essential difficulty and a possible escape from it:

> As I have already pointed out, there is an interesting problem concerning the possible interaction between the chromatin of the cells and the protoplasm during development. The visible differentiation of the embryonic cells takes place in the protoplasm. The most common genetic assumption is that the genes remain the same throughout this time. It is, however, conceivable that the genes also are building up more and more, or are changing in some way, as development proceeds in response to that part of the protoplasm in which they come to lie, and that these changes have a reciprocal influence on the protoplasm. It

may be objected that this view is incompatible with the evidence that by changing the location of cells, as in grafting experiments and in regeneration, the cells may come to differentiate in another direction. But the objection is not so serious as it may appear if the basic constitution of the gene remains always the same, the postulated additions or changes in the genes being of the same order as those that take place in the protoplasm. If the latter can change its differentiation in a new environment without losing its fundamental properties, why may not the genes also? This question is clearly beyond the range of present evidence, but as a possibility it need not be rejected. The answer, for or against such an assumption, will have to wait until evidence can be obtained from experimental investigation.

Morgan's view must have come as a shock to orthodox geneticists, for whom he had become the defender of the constancy and integrity of the gene. The dogma of nuclear constancy during development, which was a chief source of the dilemma of gene-controlled differentiation, has in the meantime been considerably qualified. Indeed, it now seems a tenable view that "the genes are changing in some way" during development, as Morgan had suggested, but actual evidence of this (*e.g.*, from nuclear transplantation experiments) was obtained only much later.

– Chapter 20 –

THE RISE OF POPULATION GENETICS

POPULATION GENETICS HAS COME to occupy a rather special place in biology. It represents the interest in the processes of evolution and in the improvement of domesticated plants and animals which strongly motivated the early students of genetics. It made possible the great renaissance of evolutionary biology which began about 1930. Population genetics then tended to reunite fields of biology such as genetics, ecology, paleontology and systematics, which had tended to take separate paths. It has thus been referred to, and with cause, as the core subject of general biology.

Population genetics deals with the frequencies and interactions of genes in interbreeding populations, with agencies such as mutation, natural and artificial selection, migrations, mixing of races and chance factors which tend to alter gene frequencies and thus to cause evolutionary changes. Its problems are stated in purely biological terms and are pursued by methods which are peculiar to genetics. Genes, and those chromosome structures which, like genes, retain their integrity over long successions of generations keep the abstract character which they had at the birth of genetics itself, that of elements which segregate without contamination, and are thus units subject to mathematical treatment.

As other genetical problems have been or eventually will be stated in molecular terms questions about them take the form given them by chemistry and physics. Population genetics, however, bids fair to retain its original character and, together with plant breeding and animal breeding, is likely to stand for many years to come as the exemplar of the original purposes and methods of genetics.

Its character is shown also in the precedence which theory took over experiment in the development of evolutionary genetics. As Wright (1960) has said:

> Evolution is something that happens to populations, and without a mathematical theory, connecting the phenomena in populations with those in individuals, there could be no very clear thinking on the subject.

The major advances in population genetics thus depended upon a rigorous and sustained application of mathematical theory and analytical methods to working out the consequences of Mendelian heredity in populations of plants, animals and men. This involved study of three kinds of problems: first, the genetical correlation among relatives of the various degrees which express the connections among the members of an interbreeding population; second, the analysis of different systems of mating—inbreeding of various degrees, assortative mating, random mating, etc.; third, analysis of evolutionary forces such as selection, mutation, migration and chance (random genetic drift) in Mendelian terms. In addition, of course, methods had to be elaborated for dealing with quantitative characters leading to the development of what has come to be called "quantitative" or "biometrical" genetics. In this brief historical account I shall deal only with what Dobzhansky (1937) has called "the rules of the physiology of populations" expressing the interaction of the evolutionary factors listed above. Since the ends in view have been the interpretation of evolutionary processes, both in natural populations and in cultivated plants and animals, a proper history of population genetics would include the development of theories of the mechanism of evolutionary change. Some of the mathematical aspects of this have been discussed by Wright (1959) and more generally by Dobzhansky, Ernst Mayr, G. G. Simpson, and others.

The development of theory concerning heredity in populations began in the nineteenth century but took a new start when heredity came to be conceived in terms of genes (Castle, 1903; Pearson, 1904; Yule, 1906). In this period, the most comprehensive theoretical treatment of Mendelism applied to populations mating at random was that of Weinberg (1908), which dealt with gene frequencies, gene interaction, and environmental variance (*cf.* p. 122). Hardy's one excursion (1908) into genetics was not followed up nor did Weinberg's work have an immediate effect.

Signs of newly awakened interest in the theoretical consequences of Mendelism appeared in 1913 when Pearl, in working out the effects of brother-sister mating, came to the erroneous conclusion that it causes no changes in the proportion of heterozygotes. Jennings (1914) and Fish (1914) independently saw the error and worked out the correct result—namely, a decline in the proportion of heterozygotes. Interest in this probably derived from the long history of observation and experiment on the effects of inbreeding during the eighteenth and nineteenth centuries and the availability of pedigree records of many breeds of livestock (Darwin, 1876; East and Jones, 1919). Pearl (1913) devised a coefficient of inbreeding based on the decrease in the number of different ancestors due to matings of relatives, but this had no relation to the hereditary process. Jennings (1916, 1917) developed a theory based on the relative rate of decrease in the proportion of heterozygotes in populations of size 2 (brother-sister and parent-offspring). He noted Pearson's and Hardy's conclusions concerning stability of genotype proportions under random mating. It was his incorrect application of this principle at one point in his paper which attracted the attention of Wright and Danforth (*vide* Jennings, 1917) and thus helped to set in

motion the succession of events which led to more general use
of equilibrium principles.

A period of more rapid development of theory began about
1917. The first two contributors to it had just begun to work
on theoretical and statistical problems suggested by applications of genetics to agriculture and to human populations.
Sewall Wright, who had studied experimental genetics under
W. E. Castle at Harvard University, had taken charge in 1915
of the experiment on the effects of inbreeding in guinea pigs
which had been carried on since 1906 by G. M. Rommel in
the Animal Husbandry Division of the U. S. Department of
Agriculture. R. A. Fisher had gone in 1917 from a fellowship
at Gonville and Caius College, Cambridge University, to become head of the statistical department of the Rothamsted
Experimental Station, the oldest agricultural research institution, at Harpenden, England.

Wright (1917) used equilibrium principles in comparing
observation with expectation, on the assumption of random
mating, in a case of color-inheritance in cattle, and in reject-
ing a one-gene hypothesis for the inheritance of blue eye color
in man (1918). These are the first instances known to me of
the use of population methods in discriminating between
genetical hypotheses. Felix Bernstein's (1924) use of equilibrium principles in reaching a correct interpretation of the
inheritance of the ABO blood group antigens in man is often
cited as the first example of this method.

Wright's first contributions to population genetics stemmed
from his interest in inbreeding effects in evolution. This had
been aroused first in 1909 by Vernon Kellogg's calling attention to the studies by the Reverend J. T. Gulick (1832–1923)
of land snails of the family Achatinellinae in the Hawaiian
Islands. Gulick, an American born of missionary parents in
Hawaii, had described (1872, 1888, 1905) the variation in
the shells of several species of this family and found what
appeared to him to be random differentiation of races in
similar habitats in different valleys. Since such snails repro-

duce by self-fertilization, Gulick ascribed the racial differentiation to the fixation of variations occurring by chance and
preserved by reproductive isolation. In 1888, he published an
important paper, "Divergent Evolution Through Cumulative
Segregation," presenting a view referred to by Alfred Russell
Wallace (1889) as one with which "the views of [both] Mr.
Darwin and myself are inconsistent."

It is an interesting parallel that when Wright had developed Gulick's idea in his concept of random genetic drift
(sometimes referred to as the "Sewall Wright effect"), it,
too, was attacked sixty years later as inconsistent with evolution by natural selection (*e.g.*, Fisher and Ford, 1950). In
reply to this criticism, Wright (1950) wrote:

> It may be noted here that the use of my name for the
> evolutionary effect of inbreeding is hardly appropriate.
> The first author to suggest that random differentiation
> among small isolated populations was something that
> must be taken into account seems to have been Gulick
> in 1872.

Obviously, random differentiation among populations, if
it must carry the name of the originator, should be recognized
as the "Gulick effect." It would thus rest on facts observed
first in nature and later verified by experiments in which
inbred lines of animals and plants, isolated by prevention of
interbreeding, diverged or drifted apart in readily observable
hereditary traits.

In a series of papers (1921–1923) Wright used his newly
invented method of "path coefficients" to deduce the consequences of Mendelian heredity under different systems of
mating. Inbreeding was conceived in terms of deviations from
the distributions of genotypes expected on equilibrium assumptions under random mating. A rigorous treatment of systems
of mating was given in which correlations among relatives
were deduced from the path coefficients connecting zygotes
and gametes. A formula was derived for computing a coeffi-

cient F which was closely related to the decrease in proportion of heterozygotes due to the degree of inbreeding and thus could serve as a coefficient of inbreeding. It was independent of gene frequency and thus applicable to mating systems of all kinds from random mating (F=0) to the complete homozygosis (F=1) toward which inbreeding tends. This led to a further generalization of the equilibrium formulation:

$$q^2 + Fq(1-q)\,AA : 2q(1-q)\,(1-F)\,Aa : 1-q^2 + Fq(1-q)\,aa$$

If F is constant from generation to generation, then the proportions of *AA, Aa* and *aa* will also be constant. If F is zero, the expression becomes that of the equilibrium formula, as had been shown by Weinberg.

F was used first in estimating the inbreeding practiced in the origination of pure breeds of livestock, and later came into general use to describe the breeding structure of populations. It was later extended to sex-linked heredity and to polyploids as well as to estimating the effect of linkage on fixation of genes under different systems of mating.

Interest in theory underlying the operation of the Mendelian mechanism thus began with the automatic consequences of self-fertilization as deduced by Mendel, led through consideration of the equilibrium to be expected under random mating (Pearson, Hardy, Weinberg) and back through departures imposed by inbreeding to more generalized views of the genetic system, in populations.

R. A. Fisher's interest in genetics arose out of his training in mathematics and statistics and his daily work as statistical consultant to biologists. Among these E. B. Ford was one of the first to bring problems of population genetics to Fisher's attention. Fisher's first paper in this field (1918) was "The Correlation Between Relatives on the Supposition of Mendelian Heredity." Using Pearson's data on human stature and other measurements, he showed that the correlations found were close to those expected on the theory of particulate rather than of blending inheritance. This reversed the reasoning of the biometricians who had tried to derive theories of inheritance from correlations of relatives. This helped to heal the breach that, in England at least, had tended to separate the two schools. Fisher's main results were a rederivation of some of Weinberg's (whose work he seems not to have known) by different methods and were extended to other conditions such as those of assortative mating. In 1922, he derived mathematical methods for the study of balanced polymorphism which had important consequences in the work of E. B. Ford and others who later studied the structure of natural populations of insects.

Fisher's chief influence on biology, however, was exerted through his *Genetical Theory of Natural Selection* (1930) with its "fundamental theorem of natural selection": "The rate of increase in fitness of any organism at any time is equal to its genetic variance in fitness at that time" (p. 35). Fisher's object was, in his words (p. 22):

> to combine certain ideas derivable from a consideration of the rates of death and reproduction of a population of organisms, with the concepts of the factorial scheme of inheritance [meaning Mendelism], so as to state the principle of natural selection in the form of a rigorous mathematical theorem, by which the rate of improvement of any species of organisms in relation to its environment is determined by its present condition.

Natural selection, in this formulation, will always tend to increase fitness, in the sense of reproductive fitness, and the course of evolution will be determined by the momentary advantage of one allele over the other.

This thesis which formalized the relation of genes to one evolutionary process, mass selection, had an immediate and wide effect on many fields of biology. This was largely because it set forth clearly the antithesis between the old theory of heredity as blending, on which Darwin had based his views of the operation of natural selection, and Mendel's proof of inheritance as particulate. Under the old view the progressive loss of variance from the population would require an enormous mutation rate to maintain the hereditary variety on which natural selection could operate. In the Mendelian view, the probability of retention of each new recessive mutation would ensure variety with a low rate of mutation. This had, of course, been implicit in the Hardy-Weinberg equilibrium formulation, but Fisher made it a cornerstone of the genetical theory of natural selection and brought it to recognition. For this reason, Fisher's book of 1930 is often taken as the beginning of the renaissance of the theory of natural selection.

In fact, however, mathematical models of the operation of selection were also provided in Wright's seminal paper of 1931, "Evolution in Mendelian Populations," and in a succession of papers by J. B. S. Haldane which were summarized in his book *The Causes of Evolution* (1932). It was rather the conjunction of these three publications and the somewhat delayed effect of an important work by Chetverikov in 1926 which gave rise to population genetics as a clearly recognizable biological field.

One of the architects of this development, J. B. S. Haldane, had come to genetics through biochemistry as a student of Gowland Hopkins at Cambridge University. It was probably such a student as Haldane who had evoked the rather plaintive remark of William Bateson (B. Bateson, 1928) that all the bright students seemed to be going into biochemistry. Haldane was, however, competent in mathematics as well as in chemistry and genetics and in a series of "Mathematical Contributions to the Theory of Natural Selection" (*cf.* Haldane, 1932) he worked out the theoretical effects of different forms and intensities of selection on the frequencies of autosomal, dominant and recessive, sex-linked and partially sex-linked genes, including estimates of equilibrium values when the introduction of new alleles by mutation was balanced by elimination through selection as in a steady-state population. This led to the first estimates of the mutation rate of a deleterious gene in a human population.

In reviewing the progress of genetics in 1938, Haldane said:

In the last 20 years a considerable body of evolution
theory based on Mendelism has arisen. The pioneer in
mathematical theory was Norton of Trinity College whose
work has been continued by Haldane, Fisher and Wright.

Actual use of H. T. J. Norton's work appears, however, to
have been made, among those who helped to build population
genetics, chiefly by Chetverikov (1926), who reproduced a
table computed by Norton for Punnett's *Mimicry in Butterflies* (1915). In this table, the effect of different intensities of
selection is estimated in terms of numbers of generations required to change the frequency of a gene from one equilibrium value to another. This seems to have been the first
systematic treatment of selection in a Mendelian population.
One of Chetverikov's conclusions (1926; Lerner, 1961) after
examining Norton's 1915 table is worth quoting:

First of all, we see that because of the effects of free crossing and selection, under the conditions of Mendelian
heredity, every, even the slightest, improvement of the
organism has a definite chance of spreading throughout
the whole mass of individuals comprising the freely crossing population [species]. Here Darwinism, in so far as
natural selection and the struggle for existence are its
characteristic features, received a completely unexpected
and powerful ally in Mendelism.

Norton's study of the rates at which mass selection produces
homozygosis of favorable dominant and recessive genes was
overshadowed by the later and more comprehensive studies of
Haldane. Haldane's determination of the equilibrium between
recurrent mutation and selection, culminating in the proof
that the loss of fitness from deleterious mutation is due to
mutation rate and not to the intensity of selection, disclosed a
basic principle which came to underlie the important concept
of mutational load in populations.

Although Fisher, Haldane, and Wright are usually cited together as having provided the theoretical basis for population genetics, their views of the evolutionary process and especially of the role of natural selection differed rather widely. Fisher assumed that direct action by natural selection—mass selection—was the controlling force. Selection exerted its effects primarily on single gene effects as illustrated by his theory of the acquisition through selection of dominance of wild-type (normal) alleles over newly arisen mutants. Haldane also dealt chiefly with causes of change in gene frequency due to separate factors such as allele mutation and selection applied to individual loci. In addition to these considerations, Wright proposed to find methods for dealing with the enormous variety due to recombination and gene interaction. This led him to the view that the decisive action of natural selection was that among partially isolated and differentiated local populations ("demes"). This came to be called "inter-deme" selection and became an integral part of his balance theory of evolution. Two factors additional to mutation and selection had to be envisioned among the causes of change in gene frequency, namely, inter-deme migration and random drift of gene frequencies due to accidents of sampling in small demes and possibly to fluctuation in other parameters such as selective value.

Wright's ascription of an important role to random factors as part of the network of pressures on gene frequency, on the other hand, produced rather violent criticism from Fisher and Ford which tended to delay, in Britain at least, the incorporation of random drift into general theory. The objections were largely based on misunderstandings of the role attributed to drift. It was evident in Wright's paper of 1931 that drift was not conceived as a substitute for natural selection but as one of the forces in a complex network of interacting factors. Indeed, the climax of Wright's paper was a general formula expressing the most probable change in gene frequency (q) per generation as:

$$\Delta q = v(1-q) - uq - m(q-q_m) + sq(1-q)$$

with probabilities for the next generation as:

$$[(1-q -\Delta q)a + (q +\Delta q)A]^{2N}$$

where v and u are mutation rates from and to gene a, m stands for migration exchange with neighboring populations having gene frequency q_m, s for selective advantage of a over its alleles, and N the effective population size, *i.e.*, the actual number of individuals contributing to the next generation.

The changes due to drift in the genetic constitution of small populations (such as local breeding groups of a species), could in themselves be non-adaptive and result in differentiation among local groups. If sufficient, this might lead toward local establishment of a different interaction system by local mass selection even under the same environmental conditions as in surrounding demes. This suggested a greater importance for intergroup selection, and it was this departure from the simpler schemes of selection within populations which was Wright's peculiar contribution and which at the same time drew fire from evolutionists accustomed to thinking in terms of direct acquisition of adaptiveness of genotypes.

Application of gene frequency as the primary measure of evolution also tended perforce to reduce the dimensions of the process that could be studied by this means. Changes in local breeding groups and in geographic races are small as compared with the changes in the taxonomic categories with which evolutionists had been largely concerned. The new views arising out of population genetics required that microevolution and macroevolution should be aspects of the same continuous process of change by gene-frequency alteration, and this encountered the resistance of some classical systematists and paleontologists.

The major features of Wright's view of the evolutionary process, which made his theoretical work the foundation on which modern evolutionary genetics was primarily based, were first, that the state of a population represents a balance among systematic and directed pressures of recurrent mutation, selection, immigration, and random fluctuations of gene frequency; and second, that changes in such a dynamic equilibrium have a trial-and-error character. Natural selection was the final arbiter and in this the view was clearly Darwinian; but it set out in orderly fashion those other parameters which came to view with the theory of the gene, all subject to one basic measurement—gene frequency. Wright's statement of this view was given in the concluding paragraph of his 1931 paper:

Evolution as a process of cumulative change depends on a proper balance of the conditions, which, at each level of organization—gene, chromosome, cell, individual, local race—make for genetic homogeneity or genetic heterogeneity of the species. While the basic factor of change—the infrequent, fortuitous, usually more or less injurious gene mutations, in themselves, appear to furnish an inadequate basis for evolution, the mechanism of cell division, with its occasional aberrations, and of nuclear fusion (at fertilization) followed at some time by reduction make it possible for a relatively small number of not too injurious mutations to provide an extensive field of actual variations. The type and rate of evolution in such a system depend on the balance among the evolutionary pressures considered here. In too small a population (1/4N much greater than u and s) there is nearly complete fixation, little variation, little effect of selection and thus a static condition modified occasionally by chance fixation of new mutations leading inevitably to degeneration and extinction. In too large a freely interbreeding population (1/4N much less than u and s) there is great variability but such a close approach to complete equilibrium of all gene frequencies that there is no evolution under static conditions. Change in conditions such as more severe selection, merely shifts all gene frequencies and for the most part reversibly, to new equilibrium points in which the population remains static as long as the new conditions persist. Such evolutionary change as occurs is an extremely slow adaptive process. In a population of intermediate size (1/4N of the order of u) there is continual random shifting of gene frequencies and a consequent shifting of selection coefficients which leads to a relatively rapid, continuing, irreversible, and largely fortuitous, but not degenerative series of changes, even under static conditions. The rate is rapid only in comparison with the preceding cases, however, being limited by mutation pressure and thus requiring periods of the order of 100,000 generations for important changes. Finally in a large population, divided and subdivided into partially isolated local races of small size, there is a continually shifting differentiation among the latter (intensified by local differences in selection but occurring under uniform and static conditions) which inevitably brings about an indefinitely continuing, irreversible, adaptive, and much more rapid evolution of the species. Complete isolation in this case, and more slowly in the preceding, originates new species differing for the most part in nonadaptive respects but is capable of initiating an adaptive radiation as well as of parallel orthogenetic lines, in accordance with the conditions. It is suggested, in conclusion, that the differing statistical situations to be expected among natural species are adequate to account for the different sorts of evolutionary processes which have been described, and that, in particular, conditions in nature are often such as to bring about the state of poise among opposing tendencies on which an indefinitely continuing evolutionary process depends.

Observation and Experiment in Population Genetics

While the theoretical basis of population genetics was being worked out, largely during the period 1917–1932, the first steps toward the study of actual populations in nature were already being taken, and relatively independently of the construction of theory. The extensive studies of Francis B. Sumner (1932) on the variation in geographic races of mice of the genus Peromyscus, and of Goldschmidt (1934) on geographic races of the gypsy moth (Lymantria) pointed the way toward the kind of studies that were needed. The interpretations given of the above sets of carefully gathered facts pointed also to the need for theoretical analysis of the consequences of Mendelian heredity since Sumner had appealed to Lamarckian explanations and Goldschmidt to adaptive mutation. In addition, naturalists and taxonomists had made extensive observations of the variations within species and races in nature, but in general their studies were not guided by the kind of rationale available in Mendelian heredity. Isolated cases showed how valuable such a rationale could be in the study of evolution, as when, in 1921, J. H. Gerould gave an example, possibly the first, of the selective elimination of a mutant gene from a butterfly population by a predator.

In 1926, one of the foundation papers of population genetics appeared. This was the essay of the Russian geneticist S. S. Chetverikov, "On Certain Aspects of the Evolutionary Process from the Viewpoint of Modern Genetics," which accurately brought into focus the essential relations between Mendelian heredity and evolutionary processes. Perhaps because it was published in Russian (an English translation appeared only in 1961) it failed to have an immediate effect in other centers of genetic research. However, it was probably responsible for the rapid rise of a vigorous movement in population genetics among Russian-trained genticists such as Dobzhansky, Dubinin, Serebrovsky, Timoféeff-Ressovsky and others.

One of the most important of Chetverikov's conclusions was that populations in nature maintain within themselves the variants which arise within them by mutation. The species, in Chetverikov's terms, should absorb mutations "like a sponge" and retain them in heterozygous condition indefinitely, thus providing a store of potential but hidden variability out of which the adaptiveness of the population to a changing environment could arise. Chetverikov (1927) supported this prediction by finding many different hereditary variants among the offspring of wild females of *Drosophila melanogaster* which had been inbred in the laboratory. Similar attribution of variety in natural populations of rodents as due to mutation had been put forward earlier (Dunn, 1921), but it rested on analogy rather than experiment. Following Chetverikov's initiative, Dubinin and others soon found that most wild Drosophila populations contained high frequencies of mutant genes, including lethal genes.

The source of the variety in natural populations of Drosophila was clearly established, since the same kind of variants which the natural populations carried concealed by their dominant wild-type alleles had already been shown to arise by mutation in laboratory stocks. This provided a stable groundwork for experimental population genetics.

A review and synthesis of the first fruits of actual observations of Mendelian populations were given in Dobzhansky's book of 1937. The title, *Genetics and the Origin of Species*, may have seemed presumptuous to those who had been unaware of what had been happening in genetics. The book, however, justified its title by showing that the mechanisms of evolution can be analyzed by genetical ideas and methods combined with observations of populations in nature and in the laboratory. This book, together with the essay of Timoféeff-Ressovsky (1938) showed how the gap between theory and observation could be closed and brought population genetics the recognition not only of geneticists but especially of biologists of all kinds. Thereafter, the basic conception of the gene pool, composed of all of the alleles, many of them differentiated into wild-type and mutant, in a Mendelian population, became the model on which evolutionary studies were based. The extent to which such ideas and methods have transformed evolutionary biology can be seen in Ernst Mayr's *Animal Species and Evolution* (1963).

The testing of theories of population genetics was carried out mainly with the species of animals and plants which had been used in analyzing the transmission mechanism and the mutation process. Man had not proved to be very useful for such analytical purposes—but it was otherwise when Mendelian populations were the units of study. The first material for human population genetics was provided by the human blood groups. In 1910, von Dungern and Hirszfeld showed that the ABO blood-group differences (the blood groups discovered by Landsteiner in 1900) were determined entirely by heredity, although their interpretation that A and B antigens on the red blood cells were caused by two independent genes was later proved to be wrong. Ludwik Hirszfeld (1884–1954) and his wife Hanka, two Polish physicians, made the first blood-group gene-frequency studies while serving with the allied armies at Salonika in the First World War. They were the first to describe (1919) racial variation in ABO blood-group frequencies and thus introduced serological genetics into anthropology. Their discovery of blood-group gene polymorphism and of regular geographic clines of changing frequency (for example, blood group B increasing from west to east across Europe and Asia) prepared the way for the most extensive genotypic description of geographic populations which is available for any species. Implicit in their work was the view that race differences can be described as differences in the relative frequencies of genes which occur in all populations.

Of importance also in the development of human population genetics was the application of equilibrium principles in the study of the inheritance of the ABO blood-group alleles by Felix Bernstein, the mathematical statistician of Göttingen (1924). Bernstein's demonstration that it was possible to derive reliable conclusions about the transmission system of heredity from the proportions of phenotyes in a random-breeding population was subsequently widely applied to other human hereditary variations and to many cases of polymorphism in animal and plant populations.

– Chapter 21 –

CONCLUDING REFLECTIONS

AT THE END of an account, however incomplete and preliminary, of the growth of some central concepts of genetics, it may be interesting to try to identify some of the stages in the hierarchy of operations through which a new science evolves. Such operations will, of course, not be peculiar to genetics. In general, however, one would expect them to be more obvious in genetics than in older fields of science. This should be so, partly because the history of genetics falls mostly in a period when the published record of research is open and readily available, and partly because the observed facts or data of classical genetics are derived chiefly from experiments and the relation of these to theory is usually readily perceived. Genetics, as many students have testified, can be learned by imitating the steps by which its principles were established, that is by deriving such rules from sets of numerical data in the form of problems, and vice versa, by predicting the outcome of experiments from a knowledge of principles. What then are some of the actual operations by which such a science progresses?

The initial event which sets a science in motion has been called "discovery." The implication is that something entirely new has been perceived. But what is it that is discovered?

It has been said that Mendel discovered the principle of segregation of alleles—as though the principle had existed and had waited to be uncovered. That was true of the proportions 1 : 2 : 1 or 3 : 1 in the offspring of hybrids, which suggested the existence of segregating elements. Such distributions had been observed and recorded by Darwin and by W. J. Spillman, to take two recorded cases. However, the disclosure of these facts did not lead to the development of genetics, while the *idea* of segregating elements did. The idea was a product of Mendel's imagination and was independently arrived at by Correns and de Vries and perhaps by others. "Inventing" or "conceiving" (in the same sense as the event which precedes birth) would perhaps more nearly describe the introduction of a new idea.

Mendel's case illustrates how difficult it is to trace the relations between fact and idea. In his paper, the theory of segregation and independent assortment of elements is presented as though inferred from the facts reported there, that is, from the proportions of phenotypes observed in the generations following a cross. Perhaps it was originally so derived. Certainly the idea can be most quickly understood by reasoning backward from facts to the explanatory assumptions. But, as we have seen, there is doubt whether it was derived from only those facts presented in the paper, since the fit of these observations to those expected on the basis of the theory is just improbably good. This suggests that the theory was a "preconceived idea," but we are left in the dark as to the time or manner of its conception. A man more given than Mendel was to recording what went on in his mind might have left us some clue as Darwin did when he described how the idea of natural selection struck him after reading Malthus on overpopulation. Even though we cannot specify the details of the process and however we refer to it, the origination of ideas lies at the first level.

Ideas in this sense may take the form of hypotheses, suggested explanations, or theories. Mendel referred to his as theory, law, or principle, terms which are differentiated from the first form of the idea by having evidence in support of them. One has the impression that, since Mendel had set out to find a "generally applicable law governing the formation and development of hybrids," if he had failed to find such a law he would not have mentioned any lesser category of idea, such as an hypothesis that had not been sustained.

The normal growth of science, however, requires choosing between alternative hypotheses and the discarding of the wrong or less probable one. This process is often referred to as "proof." If we say that Bridges proved, by his work on nondisjunction, that genes are physical parts of chromosomes, we mean that the outcome of his observations and experiments were as a whole more consonant with that idea than with its opposite. An alternative, that some elements like genes are not parts of chromosomes, was not thereby disproved, and the continued reference to plasmagenes and "non-chromosomal" genes testifies to the importance of defining terms, in the context and for the purposes for which they are used. We say that "operational definitions" are better adapted to the kinds of change which genetics must undergo than definitions which become fixed and thus restrictive. "Proof" may come from experimental tests of hypotheses or from a weight of observational evidence, as in the case of natural selection. It is clearly a second-order operation, not primary as in the origination of ideas, but in genetics it has often gained more credit and in general has had a higher status than "theorizing."

Another discernible operation in the advancement of genetics has been what I can only refer to as "calling attention to." This was of greater importance in the early stages of development of the science than it is today, since genetics had to start from scratch in 1900. It grew by proselytizing adherents from older fields of biology. The first generation of those we now recognize as geneticists were, before 1900, something else. Bateson, Castle, Davenport, Morgan, and Wilson came from embryology; Correns, de Vries, and Tschermak, from plant physiology; Cuénot, from anatomy and physiology; Shull, from ecology and taxonomy. Two of these, Bateson and

Davenport, especially were distinguished by their influence in attracting others to the study of genetics in its earliest stages. Castle, a protégé of Davenport's, served the same function. All spoke to wide and general audiences of both scientists and laymen. It was less propaganda than exposition that they practiced. Bateson especially quickly envisioned the role that genetics was to play in the interpretation of evolution and development, described its problems, and provoked the controversies that gained first adversaries and then adherents— both useful in building a new science.

It is less easy to identify and describe an activity that in a sense serves to hold all this together and to weld it into the intellectual unity which is especially marked in genetics.

In the chapter entitled "Clarifying Ideas" I have described the influence of Johannsen as critic and orderer of ideas. The application of logic to ideas is accompanied by, and springs from, conceiving them in abstract, symbolic terms. Galton's strength and influence derived from this facility, and it is no accident that it was Galton who founded biometry and the statistical study of variation, nor that it was Johannsen who first gave such methods currency and status in the new genetics and reconciled the old fact of continuous variation with the new idea of discontinuous origins of variation by mutation. Johannsen as the systematizer of thought in genetics later came to defend his brain child, the gene, as pure symbol and refused to allow it corporeal existence. H. S. Jennings was another such orderer. It was Jennings who examined the logical basis of the chromosome theory of the Morgan school and helped to give it currency in general biology. He, too, was attracted to the mathematical formulation of genetical processes such as the decline of heterozygosis under inbreeding. Wilhelm Weinberg was certainly another such orderer and analyzer but as a practicing physician, self-taught in mathematics and limited to evening hours at an overburdened and untidy desk, he produced rather tortuous papers that did not exert an immediate influence. Karl Pearson, on the other hand, although remaining skeptical of the basic assumptions of

Mendelism, led in the preparation and organization of the statistical methods upon which genetics became increasingly dependent. In this he was preceded by Galton and followed by R. A. Fisher. Fisher's recognition of the nature of Mendelian heredity and his inventiveness in mathematical statistics and in the design of experiments brought his methods into general use as a critical language for genetics. The chief clarifications in the application of genetical ideas to the processes of evolution came from Sewall Wright, with important contributions from Fisher and J. B. S. Haldane. The wide range of factors brought to bear on the processes of change in populations, in Wright's balance theory, had a major effect not only on genetics but also on the unification of biology as a whole. In this development, the ordering influence of the kind of thinking characteristic of genetics came to be recognized even in those parts of biology governed by descriptive and historical methods, that is, those which had seemed most distant from theoretical genetics.

Discovery and Rediscovery

The growth of genetics illustrates also another feature which appears to be characteristic of modern science. This is the tendency, which Robert K. Merton (1961) considers a rule, for the same discovery to be made independently by different investigators. Merton has given evidence for his view that "singletons" or unique discoveries are exceptions and may appear to be unique only because their prior publication forestalled a later announcement of the same discovery made by another scientist.

The history of genetics provides some prime examples of this. There were, after all, three published rediscoveries of Mendel's principle. One of these (Correns') would not have been published when it was, had not de Vries' paper come out · first. Correns did not regard his own discovery, in the late 1890s, as equal to Mendel's in the 1860s and intended to withhold his until new points, not touched by Mendel, had been made clear. Correns' discovery could thus, under certain circumstances, have become what Merton calls a "forestalled multiple," leaving the original (as far as Correns was concerned) to stand alone, *i.e.*, a singleton.

Again, what might be called Garrod's principle of a relation between a gene and a specific metabolic reaction, was rediscovered by Beadle and Tatum some thirty years later, after which biochemical genetics developed with increasing rapidity. The idea as it appeared after knowledge of the structure and functions of enzymes had grown, in the "one gene, one enzyme" and "one gene, one polypeptide" theories, was, of course, different in content from what it had been in the period before this knowledge existed. The essential continuity or repetition was in the form of reasoning employed, from defect in a reaction to alteration in a gene and vice versa.

The several separated announcements of a chromosome theory of heredity extending over a period of thirty years, say from 1884–1885 (*cf.* p. 108) to Bridges' proof of 1916, do not fit the pattern of discovery and rediscovery as in the earlier examples. There was in the first place no "discovery" of a single idea except in a very general sense, and the theories of Weismann, Boveri and Sutton while bearing similar names had differing contents. Correns had the idea of genes in a kind of linear order in the chromosome long before Sturtevant provided proof of it, but the idea in the form Correns gave to it could not have become the principle that emerged from the Drosophila experiments.

A last example reveals the fundamental difficulty in classifying "rediscovered" ideas as independent. This concerns the famous coincidence of the almost simultaneous publication, in 1908, of the generalized form of Mendelian equilibrium by Hardy and by Weinberg. There is no evidence that either knew of the work of the other, but certainly both of them knew of Mendel's work and of Pearson's more limited generalization. They were connected, as it were, through common ancestors in the field of ideas, and this is a limitation on "independence" which is often difficult or impossible to evaluate. The "idea" in this case was a further extension of reasoning in algebraic form which had already been introduced by Mendel. The equilibrium idea had great importance for biology. For a mathematician like Hardy, however, to generalize such a relation was a passing application of a mathematical operation, which might have served as recreation. Punnett in his account (1950) says that after telling Hardy of the problem, Hardy "replied with pr = q²."

The interest in such cases is not in questions of personal prestige or priority of discovery. These have seldom appeared in genetics. It is, rather, the interpretation which replicate discoveries, separated in time, suggest for the periods of stasis or delay with which the progress of science is punctuated. There was of course that period from 1865 to 1900 when Mendelian genetics stood still. Biochemical genetics resumed its interrupted progress in 1936; population genetics took off on a new start after 1920; evolutionary genetics, after 1930. The chromosome theory in its developed form was dormant from 1904 until 1911. Human genetics could have got under way in 1902, or certainly following Weinberg's lead in 1908, but in fact made little progress until thirty to forty more years had passed.

Rediscovery provides one of the explanations for resumption of progress. In a sense, it is like recurrent mutation which provides repeated opportunities for the utilization of something new under conditions which differ somewhat at each recurrence.

But that is not all that can be said about the periods of delay in the evolution of scientific ideas. Some attention must be paid to the resistance on the part of scientists to new ideas. Thus, Mendelian genetics was at first slow not only in its own development but also in penetrating other fields of biology. Resistance to genetical ideas was not merely stubborn or willful but often due to honest doubts fostered by experience which had not required grappling with such ideas.

Thus, in the early days of genetics what was often lacking was a clear understanding of the nature of Mendel's primary assumption. The idea that the elements of heredity are highly stable and not subject to fluctuating variability was repugnant to many biologists. These included for a time William Bateson, W. E. Castle, T. H. Morgan, and others who helped to build the new science. There had been a natural growth in nineteenth-century biology of faith in the opposite assumption: namely, that biological forms and properties were inevitably subject to variation. The closer the biologist had been to Darwin's ideas and evidence on variation as the condition of evolutionary change, the more firmly did he hold this faith.

W. E. Castle was a conspicuous example of those who held the view that genes must be modifiable by selection. It was shared by many others to whom the inviolability of the gene to change from its genotypic environment in the heterozygous state seemed like arbitrary dogma. Castle cured himself of disbelief in the integrity of the gene the hard way—by 15 years of arduous experimentation.

Belief in the possibility of the inheritance of acquired characters survived in the face of lack of evidence long after Mendel's principle and the better knowledge of mutation had made such an assumption generally unnecessary. This idea seemed to persist longest in areas more and more distant from experimental Mendelian genetics, for example among some embryologists and psychologists and especially in philosophical doctrines such as holism. The fact of principal importance in the present context is that it has long since ceased to play a role in determining the course of thought or observation in genetics. The earlier history of this idea and its influence on the study of heredity has been well described by Conway Zirkle (1946).

It must not be thought that the effects of failure to appreciate the implications of a new idea, such as that of the integrity of the gene, were primarily of a negative kind. It is true that this was one of the causes of delay in the development of genetical ideas. It was also the cause of progress since it led to observation and experiment which elucidated other areas of genetics. The important effect of Castle's experiments on selection, for example, was to give strong support to the multiple-factor theory to which he had been opposed as the basis of theories of selection. When we recall, too, that it was T. H. Morgan's doubts about the integrity of the gene ("Once crossed, always contaminated," he had said in 1905) which first brought him into genetics, the lack of immediate appreciation of a new idea can be seen in its true light as one of the primary elements in the advancement of knowledge. After a new idea has served its main function as a launching platform for further investigation, we tend both to forget the doubts and opposition which it had first engendered, and at the same time we come to regard it as established, hence "right" and "correct."

However, we should hardly use such words today to describe the view of the gene as generally conceived in the first decade of this century. It has not proved to be invariable or inviolable and has been "contaminated" by having its parts broken and recombined in hybrids. It has become subject to deliberate and controlled modification by a variety of mutagenic agents, and has become altogether a more complex concept than was the gene of the first chromosome maps. We know something now about the conditions for its relative stability in transmission; and we owe the opportunity for acquiring our present view (which seems more sophisticated to us) to the earlier and cruder concept of "unit factors." The defense of that view to which the early Mendelians devoted so much effort had its chief outcome not in proof of the rightness of the idea in that early form but in the better form which serves the needs of genetics today.

It would unduly prolong these reflections (and the length of what was intended to be a "brief account") to summarize in detail the changes in each of the main lines of thought which were outlined in the introduction. It will serve, however, to emphasize a main point, that is, the rather impressive unity which genetics has retained, if we consider anew the five "lines," now in relation to each other.

The main line is clearly that which led from Kölreuter

through the experimental plant hybridizers of the late eighteenth and the nineteenth centuries and thus to Mendel, or better, to the quartet which included also de Vries, Correns, and Tschermak. The main idea was that heredity is particulate, and the chief generalization was what has been called the theory of the statistical gene, in which the gene as an element is inferred from the distribution of differentiating characters in the offspring of hybrids. This was the kind of idea that could and did arise out of experience in plant breeding and later in breeding experiments with animals. It is noteworthy that when T. H. Morgan (1926, p. 32) summarized the theory of the gene in one sentence, the five principles he formulated (segregation, independent assortment, crossing over, linear order, and limitation of linkage groups) were "derived from purely numerical data without respect to any known or assumed changes in the animal or plant that bring about in the way postulated, the distribution of the genes." This was the apex of the main line, a coherent, logical system without reference to a physical basis. The first two principles constituted what had been called "Mendelism." It can now be seen as the central focus through which passed those other genetical ideas which had been developing in the nineteenth century. They concerned the physical basis of heredity, the origin of hereditary variations, the mechanism of evolutionary change and living units as speculative constructs by which development, regeneration, metabolism, and (incidentally) transmission, and hence heredity, were to be explained.

As these passed through the focus of Mendelism they were changed as light rays are when passing through a lens. As the ideas emerged from this transformation in the early years of the twentieth century, they became respectively: the theory of the gene as the basis of formal or transmission genetics, the chromosome theory of heredity as the basis of the pattern or architecture of the hereditary material, the idea of mutation as the primary cause of hereditary variety through misreplication of particles, changes in gene frequency in populations as the essential feature of evolution, and the dynamic or physiological view of the gene as the mediator and modulator of steps in sequential processes in metabolism and ontogeny.

All of these were related to each other through those few focal ideas which derived chiefly from the main line, that is, the theory of the gene. Their subsequent development was influenced mainly by the habit of thinking in terms of experimentally demonstrable, self-replicating elements of high, though never absolute, stability. Theories based on speculatively conceived living units, such as pangenesis, withered and died after 1900. They were no longer useful and never did pass through that focus established by Mendelism. The main line continued in plant and animal breeding, in population theory and evolutionary genetics, and in more refined theories of structure of the hereditary material which were based on studies of mutation and on that most characteristic operation, breeding analysis.

One can identify in this work and this period (1900–1929) what can be called "purely genetical" methods of study, which came to be trusted for developing genetical theories more fully than other forms of observation. What Morgan referred to as "purely numerical data" generally seemed to lead the way.

One of the reasons for the rapid growth of genetics in the period 1900–1920 was that the collection of data from breeding experiments could be quickly learned and had a wide appeal. That this constituted the chief activity in genetics is shown by the fact that a large proportion of the published record of that period is devoted to the detailed accounts of such experiments. These were the "data papers" that carried high rank and had major influence in establishing central principles. For example, the linear order in a linkage group could be set forth as a consistent system *before* the analogous structures in the chromosomes were known. The same reasoning which led to the elucidation of the architecture of the genetic system in higher organisms now serves as the basis for an analogous investigation of the order in which genetic elements are transmitted from one cell to another in bacterial conjugation or by transducing phages in which the structural analogues of chromosomes remain hypothetical. The fruitfulness of these ideas lies in the logical consistency of the observational data rather than in the degree of verification of the hypothesis by reference to structural elements.

Another indication of the primacy of what we have referred to as "purely genetical" methods of study is the temporal order in which present concepts of biochemical genetics appeared. The proof of gene-controlled single steps in sequences of biosynthesis appeared before, not after, the substances participating in the synthesis had been chemically identified. It was the experimental proof that a diffusible substance in Ephestia and Drosophila was controlled by a specific gene locus that led the way toward the resolution of reaction sequences into separate gene-controlled steps. The logical structure of the theories elaborated by work on microorganisms was based on ideas first made plausible by observations on higher animals and plants.

One should not carry too far this insistence on the primacy of purely genetical observation. In the first place, it represents application of operational procedures in the sense of Bridgeman rather than something peculiar to genetics. L. J. Stadler (1954) has shown how useful such methods are in defining such a central concept as the gene. In the second place, feedback relations between genetical hypotheses and observations at other levels (using microscopy and physical and chemical methods) arise as soon as first verifications in terms of structure are obtained, so that subsequently primacy, in the sense of controlling influence on the development of ideas, passes back and forth between the purely genetical and the structural observations. This modification and expansion of methods became especially marked in the subsequent development of biochemical genetics and was responsible for the revolution referred to in the following chapter, which was carried out largely by biochemists.

It is useful to recall, finally, that in genetics as in biology, generally, both the thinking and the physical operations of study proceed on different levels, and it is chiefly the appropriateness of the methods to the particular level of organization which is important.

Genetics, at its beginning, dealt with whole organisms: the higher plants and multicellular animals. The analysis of the transmission mechanism often referred to as formal genetics was the outcome of manipulations at this level. Above this lay the population composed of interbreeding individuals which has its own rules, what Dobzhansky called "the physiology of populations"; below it lies the cellular level in which the interactions and functioning of genes take place as dealt with in physiological and developmental genetics and in the cytological basis as disclosed by cytogenetics. At the lowest level lie the molecules of DNA as organized in genes and chromosomes or analogous structures. It is, of course, at this level that the process occurs which is responsible for all the levels above, *i.e.* self-replication or duplication of macromolecules with its inevitable accompaniment of misreplication or mutation.

Genetics, beginning with the middle level–individuals–moved first to form population genetics, evolutionary genetics, ecological genetics and the like, and then progressively, to levels of chromosomes, genes, and molecules.

It is unfortunate that in discriminating levels of organization we must perforce use terms such as "higher" and "lower," for this seems to suggest that the lower levels are more basic or more important. We should recall, however, that in our own species, the lives of all members are dominated by interactions at levels not yet dealt with by genetics–the behavioral, social and intersocial–and that understanding of evolutionary processes, a central problem of biology, and in fact biological problems generally, now requires cooperation of studies at all levels.

This movement of genetics in the direction of other levels has continued of course and has become more marked in the years since 1939. Some of the aspects of the earlier formative period may be made more distinct by comparison with the present. I have therefore added a chapter which may serve as a kind of bridge between classical (or organismal) and molecular genetics.

The book as first planned, however, ends here. I cannot close without recalling a reply which I feel sure many a degree candidate has given, at the close of an oral examination on his dissertation, to the question, "What did you learn from all this work?"

"How to do it better the next time!"

– Chapter 22 –

POSTLUDE: OLD AND NEW IN GENETICS[23]

MANY PERIODS OF CHANGE in scientific knowledge have been referred to as revolutions. In biology the major change was often cited as the "Darwinian revolution." In the succession of events leading to genetics, the "Mendelian revolution" was primary. Its origin was contemporary with the *Origin of Species,* but its effects were first felt forty years later, beginning in 1900. The development and generalization of the chromosome theory of heredity, led by T. H. Morgan and his associates, occurred in a short space of years, the climax of which was marked by the publication of *The Theory of the Gene* in 1926. In Darwin's case the threat to established authority and the bitter controversies provoked by his book of 1859 gave a radical and revolutionary flavor to a doctrine of evolution by slow change. But in the case of Mendelism after 1900 and the Morgan school (1910–1920), it was the rapidity with which new ideas were developed and extended which marked the change. The new was added to the old.

[23] This chapter is a somewhat abridged version of a paper written for the May, 1964, *Bulletin of the New York Academy of Medicine.* The issue was devoted to genetics, and it seemed appropriate to compare the "classical" period up to 1939, to which this book is devoted, with the new or "molecular" genetics, which took form in the 1950s. It may thus serve as a kind of bridge between the present state of genetics and the state of the ideas prevailing at the end of the earlier period.

This hardly constitutes revolution, although in biological circles it was accompanied by the kind of excitement which is evoked by social and political upheaval. In science, such excitement finds its outlet in new research which increases in volume and scope. This feedback effect enhances the speed of change.

We are now in the midst of a new period of very rapid increase of knowledge of the mechanism of heredity, and it is accompanied by the intense interest and excitement of new discovery. Already we hear the cracking of the genetic code referred to as "revolutionary." The prospect has been opened of viewing the processes of heredity on a molecular level, certainly in a more general and elementary way than had ever been possible before. Even though "revolution" may not be the appropriate word for this or any other scientific development, still its use should cause us to reflect on the nature of the change which is now occurring. Will established theories of genetics be destroyed? Will the new knowledge cause a radical change in our perception of the problems of genetics as these had taken form in the "classical" period? Does the "new genetics" deal with new questions, and can the new be added to the old, as happened in the earlier periods of rapid change?

It will help, in the consideration of these questions, if we keep in mind what have been the chief directions of research in classical genetics. In broad terms, these have dealt first with the transmission mechanism of heredity and the physical constitution of the genetic material; second, with heredity and variation (mutation) as these operate in populations, and thus with the causes and mechanism of evolution; and third, with the functioning of the genetic material in controlling the metabolism and development or morphogenesis of the individual. These are usually discussed under the rubrics of formal genetics (or transmission genetics), population genetics, and physiological or developmental genetics. These· are obviously distinctions of convenience for purposes of study by different methods adapted to the analysis of different questions; but it should be clear that all seek to understand a single entity, the succession of living organisms in reproduction, heredity, evolution, and individual development.

The chief focus through which these problems came to be viewed was the concept of the gene. This was a statistical concept resting at first simply on the occurrence of Mendelian segregation and assortment, later becoming an element in a logical system of linkage groups, inferred from breeding analysis, subsequently to be associated with individual chromosomes by cytogenetic methods in the 1920s and 1930s. Although some cases of transmission by non-chromosomal elements were known, it was concluded that, by and large, the problem of the transmission system of heredity had been solved when it had been shown to consist of genes which could be assigned to a linear system of bead-like loci in the chromosomes.

It could, in fact, be said that the problem of the transmission system has been solved three times: once by the discovery of the statistical gene, once by the proof of the beads-on-a-string theory, and now by the proof that the genetical material consists of a linear sequence of base-pairs, the order in which these are arrayed in a polynucleotide corresponding to a specific instruction for the incorporation of an amino acid into a part of a polypeptide chain of a protein. Each claim to a solution is justified, for each had answered the question which had been asked. Progress came from the ability to ask new questions and from the development of new methods.

The "middle solution," i.e., the theory of the gene, as stated by Morgan (1926) was explicitly restricted to the transmission system. But it was obvious that the constitution of the genetic transmission system could not be understood except as a result of a process of evolution guided by natural selection. The elements of heredity, whether they be considered as single nucleotides or as higher orders of pattern in the arrangement of such elements (genes, cistrons) are not merely parts of individual organisms but of continuous lines of descent which constitute the populations in both time and space from which races, species, and higher taxonomic categories evolve. The thousands of separable elements could appear in a huge number of combinations, and the action of natural selection would lead to the disproportionate multiplication and transmission of those combinations which were internally harmonious and adjusted to their environments.

This view of the genetic system as a part of a continuum led, as soon as the gene concept took form, to two essential questions: how does the gene reproduce itself, and how do new and different gene forms arise? Classical genetics was unable to give more than a formal description of the first process. What it learned about gene reproduction came chiefly from experimental studies of the second question, that of mutation, initiated by de Vries at the end of the nineteenth century and carried forward with great brilliance and persistence chiefly by H. J. Muller. In the year following Muller's demonstration of artificial transmutation of the gene (1927), a new clue appeared from an unsuspected source, Griffith's discovery of the transfer of specific transforming material from one bacterium to another. When Avery, MacLeod, and McCarty proved (1944) that the material transferred was deoxyribonucleic acid, thereby identified as the specific hereditary material capable of transmitting a new hereditary property without sexual crossing, then a new kind of genetics began.

It was proper to call this "molecular genetics" since questions about continuity and change could now be asked in molecular terms. Specifically, they became problems of DNA structure, transmission, and function in producing the hereditary characters of organisms. This substance, usually in association with protein, is found in chromosomes wherever they occur, and has proved to be the transmitter of genetic specificity from generation to generation in all organisms from virus to man, except for a few plant viruses in which the genetic material appears to be RNA. It soon became evident why a nucleoprotein should form the universal basis of heredity: it was able to reproduce its own molecular structure. It thus possessed the essential property of self-reproduction.

In the meantime, beginning in the late 1930s, the horizons of classical genetics, which had until then been largely confined to cross-fertilizing plants and animals, including man, were greatly expanded. In filamentous fungi, yeasts, bacteria, and then viruses were found systems of genetical transmission which could be resolved into elements like genes. These could enter into recombination by methods of reproduction supplementing and sometimes substituting for sexual crossing. It now became possible to test some ideas which had appeared earlier on new and more favorable experimental materials.

Garrod, in the period following the rediscovery of Mendel's principles, had entertained the idea that genes, causing the kinds of diseases which he referred to as "inborn errors of metabolism," produced their effects by blocking specific chemical steps in reaction sequences. The blocks were suspected of being due to failure of gene-operated enzymes. Such a conception was rediscovered in a gene-controlled sequence in Drosophila, but it was the rapidly reproducing bread molds, growing on chemically defined media, which enabled Beadle and Tatum and others to put the "one gene, one enzyme" hypothesis on its feet, and it was then rapidly exploited in other microorganisms. This led to the view that the primary function of a gene was to impart a specific structure to an enzyme, thus endowing it with a peculiar ability to catalyze one step in a reaction sequence. The old genetics had rather suspected that genes were themselves enzymes and thus proteins; now, the threshold of a new problem came to view, how genes, composed not of protein but of nucleic acid, could govern and determine protein synthesis.

In 1949, applications of methods of physical chemistry directly to the study of a protein produced by a mutated gene led Linus Pauling, H. A. Itano, S. W. Singer, and I. C. Wells to identify the specific change in the protein brought about by the gene. The discovery of the first of the abnormal human hemoglobins which they described as causing a "molecular disease"—sickle-cell anemia—was followed by the identification of a large number of other proteins, each of which owed its difference from normal structure to a mutated gene. In 1957, Ingram showed that the change due to the mutation, in the case of each of two abnormal hemoglobins, was confined to a single amino acid residue at one point in one of the polypeptide chains composing the globin. There could be no doubt that genes controlled protein structure by specifying the sequence of amino acid residues in the polypeptide chains. The assumed basic functional correspondence was then altered from "one gene, one enzyme" to "one gene, one polypeptide."

A major step in the creation of molecular genetics was the proposal by James Watson and F. H. C. Crick, in 1953, of a model of the molecular structure and manner of replication of DNA. In 1956, the pathway leading to the biosynthesis of DNA was discovered by S. Kornberg and his associates who showed how the specific polynucleotide chains of DNA are formed in the cell. This was followed by the solution in broad outline of the so-called "coding" problem—the problem of how the structure of specific DNA molecules is transmitted to descendant molecules and how this structure is translated into specific protein structure in the descendant cells.

All of this will be easily recognized as a part of the new genetics. In molecular terms, the gene, which had been conceived as an active unitary particle, now was viewed as a segment of a large number, possibly hundreds, of base pairs in a chain of thousands of nucleotides distributed in linear order in a chromosome or comparable structure. This resolution of genes, as units controlling specific functions, into linear sequences of subelements, had also been achieved by the usual methods of classical genetics—breeding experiments often accompanied by cytological examination. This sort of "fine-structure analysis" began when some of the genes of Drosophila and maize were shown to consist of subelements which in rare instances separated by a recombination (crossing over) within the gene-locus. Detecting such rare events required observations on millions rather than on the hundreds or thousands of offspring which had been sufficient for detecting intergenic recombination and for the construction of chromosome maps. These huge populations were available in microorganisms, and the climax of resolution of a gene-locus into its parts was reached in one region (cistron) composing perhaps 1/100 of the genetic map of a bacterial virus, in which Seymour Benzer and his co-workers demonstrated a linear order of hundreds of separately detectable sites of mutation. The gene, as Benzer pointed out, had ceased to have its old meaning when applied to the level of fine structure, although it still had meaning when used as the unit which could specify an enzyme.

Here we return for a moment to the assumed structure of DNA. Each of its two complementary chains was conceived as consisting of a succession of nucleotides, each consisting of a nitrogen-containing base—a pyrimidine or a purine, a pentose sugar, and phosphoric acid. The peculiar and pregnant feature of the model was the relationship between the bases, a purine on one chain, A (adenine) always being bound by a hydrogen bond to a specific pyrimidine T (thymine) on the other and G (guanine) always bound to C (cytosine). The order of the bases, *e.g.* A C G T or any variant of the order of these four elements in one strand, was to specify its own genetic effect, *i.e.* the instructions or information it was to transmit to the cells in which it occurred, particularly what kinds of enzymes they were to produce. It was also to determine the replication of this order when, at chromosome replication or copying, a daughter strand was assembled out of compounds available in the cell. Since an A base would always choose a T base as its partner, and C associate with G, the result would be the construction of a complementary linear sequence in the daughter strand, thus preserving the order of bases. This order would be the feature determining the continuity of genetic properties, that is, the essence of heredity. The kinds of departures from this order which might arise by mistakes in replication were suggested by the discovery that the change in a protein caused by a mutation was the change in one amino acid residue out of hundreds. If it should prove that the incorporation of each of the twenty amino acids which enter into a protein is governed by some specific sequence of base pairs in the DNA molecule, and that mutation is essentially a change in this sequence, then both maintenance and change of biological properties in reproduction would find an explanation. This, in outline, is what has happened in the last three years. The order of incorporation of amino acids into protein appears to be specified by a code of three-letter "words" representing the order of base pairs in DNA. The three-letter or triplet code for specifying each of the twenty amino acids is now reasonably well settled (cf. Lanni, 1964). Certainly the general principle is clear. Self-reproduction, the essence of heredity, is the copying of a code of four kinds of nucleotide. The elementary step in mutation, essential for the origination of the variety on which natural selection and other evolutionary forces act, probably results from the miscopying or mispairing of two of the usually complementary pairs, so that a change in base order in a daughter strand results. I have, of course, omitted all reference to the manner in which the code instructions in DNA are transferred via other polynucleotides (RNA) to the ribosomes where enzymes are assembled.

The functions of each enzyme depend precisely upon the arrangement of the many parts of the molecule which would thus be subject to alteration by amino acid substitutions of the sort that could occur in mispairing in DNA replication. This might occur at any point in the polynucleotide chain of DNA, and recombination or crossing over between any two adjacent nucleotides might also occur during replication. The work with bacterial viruses shows that mutations can be induced by agents which act on the bases of DNA. The thousands of mutations, both spontaneous and induced, can be localized on a linear map, in which segments having a unit function each consist of hundreds of sites separable by mutation and recombination. This kind of genetic analysis was developed out of the classical concepts and applied with modern refinements to organisms known to consist of DNA (*i.e.*, viruses). It thus formed a bridge connecting the old with the new in which the analyses of DNA and its relation to protein synthesis have been carried on by chemical and physical procedures often and most profitably in living systems extracted from and devoid of cellular structure.

It is evident now that the two kinds of study, one beginning with the organism, the other with the hereditary material, are going to yield the same kind of picture of heredity and variation. Far from revealing incompatibilities between the traditional and the molecular views of heredity, it seems that work at both levels can now be utilized in attacking the great unsolved problems of evolution and development. The practical benefits of both the new and the old have already been demonstrated in agriculture and medicine, and the tempo of progress in application will certainly now increase as confidence in the empirical knowledge of the genetical transmission system is reinforced as it has been recently.

One contrast between old and new is especially striking although it is not confined to genetics but is now characteristic of "new" science. There was an interval of some forty years between the first publication (1866) of the evidence for the particulate nature of hereditary transmission and its confirmation and extension to bisexual animals and plants generally. Another thirty years passed before it began, in 1932, to affect general biological theories like that of evolution by natural selection; and even after twenty-five more years it had not been reconciled with the basic facts of morphogenesis. But now only twelve years have elapsed since the first proposal of the model of DNA, and already the code of instructions to descendants which it embodies is recognized as a universal rule not only for inheritance but also for the structure and functioning of living matter.

The older genetics had begun to serve as a focus through which diverse biological problems could be examined and restudied. The new genetics promises to greatly increase the speed with which this will occur over a wider area, which will include parts of chemistry and physics as well as biology. It is especially notable that the same methods and attitudes have been successful in dealing with living and with non-living systems. This of course had been shown long ago by the success of general physiology and later by biochemistry, but those sciences had dealt mainly with the stationary or maintenance processes of life. Now it is clear that the advancing processes which are peculiar to living organisms—reproduction, heredity, and evolution—can be dealt with in the same way. One would like to add to this list the problems of the growth and differentiation of the living individual out of a single cell, but at present this would represent a statement of faith rather than accomplishment.

— Glossary —

aleurone   The outer flowery layer of the endosperm of a seed.
allele (or *allelomorph*)   One of two or more alternative forms of a
   gene.
allelomorph (See *allele*)
anlage (German)   Predisposition; as used in discussions of heredity, it originally meant "hereditary factor" and was subsequently
   replaced by "gene." In embryology it means the rudiment or pri-
   mordium from which a part of the organism develops.

back cross   The mating of a hybrid or heterozygote with one of its
   parent varieties.
biometry   Statistics applied to biological problems.
brachydactyly   Abnormal shortness of fingers or toes.
bud sport   A somatic mutation expressed in a bud or branch.

chiasma   In cytology, a cross figure indicating the point at which
   paired chromatids have exchanged segments.
chimaera   A plant (more rarely, an animal) composed of tissues
   of two or more different genotypes, owing to somatic mutation,
   somatic segregation or grafting.
chromosome   One of the bodies into which the chromatin (stain-
   able genetic material; deoxyribonucleic acid) of the nucleus is
   organized.
complementary interaction   The production, by interaction of two
   or more genes, of effects distinct from those produced by any one
   of them separately.
crossing over   The exchange of corresponding segments between
   chromatids of homologous chromosomes.
cytology   A branch of biology dealing with the structure and function of cells.

deme   An evolutionary unit of population defined as a local interbreeding group.

digametic Forming two kinds of gametes, *e.g.*, capable of producing sperm with an X or with a Y chromosome.

diploid The zygotic (2*n*) number of chromosomes, as opposed to the gametic, or haploid (*n*), number; also, a plant or animal with the diploid set of chromosomes.

discontinuous variation In evolutionary biology, the occurrence of gaps or sharp differences in a series of related organisms as opposed to intergrading or continuous variation, *e.g.* in persons of normal stature.

Drosophila A genus of dipteran (two-winged) flies; very widely used in genetical research.

epistasis The interaction of nonallelic genes by which the effects of a single gene difference (such as pigment inhibition) mask the effects of other gene differences (such as those affecting the kind of pigment). The latter are said to be "hypostatic."

gamete A reproductive cell, such as an egg or sperm.

gene As originally defined by Johannsen (*cf.*, p. 93): a particle having the Mendelian properties of segregation and recombinability, *i.e.* an element invented to explain breeding data. Subsequently, other properties were added: an element in a chromosome distinguishable from others by crossing over or separate mutability; an inherited unit, responsible for initiating a physiological or chemical reaction, such as synthesis of the prosthetic group of an enzyme. At present, discussions of properties to be explained are more useful than attempts at rigid definition.

genetic polymorphism The regular occurrence, in the same population, of two or more discontinuous variants or genotypes, in frequencies which cannot be accounted for by recurrent mutation.

genotype The genetic constitution of an individual.

germ plasm The hereditary materials transmitted to the offspring through the germ cells.

heterogenesis A term, no longer in common use, expressing the idea of mutation or discontinuity in the origin of hereditary variation.

heterozygote A diploid individual derived from the union of gametes bearing different alleles; *e.g.* A + a → Aa. Usually refers to specific, known gene differences.

homologous chromosomes Members of the same pair of chromosomes, *i.e.* those which pair at meiosis.

homozygote A diploid individual derived from the union of gametes bearing identical alleles, *e.g.* A + A → AA. Usually refers to specific genes such as homozygous AA, BB, and so on.

hypostasis (See *epistasis*)

karyokinesis (See *mitosis*)

law of segregation Mendel's rule that members of a pair of characters, each member having come from one parent (A + a → Aa), again separate—when germ cells are formed—into discrete gametes (Aa → $\frac{1}{2}$a$\frac{1}{2}$A). The rule also applies to separation of members of pairs of chromosomes at meiosis.

Lysenkoism The doctrines, mainly claims for the inheritance of acquired characters, promulgated in the Soviet Union by Trofim Lysenko between 1932 and 1965.

meiosis Two special cell-divisions, preceding the formation of gametes, in which the homologous chromosomes pair, replicate once, and undergo segregation so that each gamete gets *one* representative of each chromosome set, *i.e.* the 2n (zygote) number is reduced to the n gametic number.

Mendelian character A differentiating character which is inherited according to Mendel's law of segregation.

Mendelize To undergo Mendelian segregation.

mitosis (also *karyokinesis*) The process by which each chromosome replicates and divides, followed by division of the nucleus into two, each new nucleus containing a full set of chromosomes.

mutation A sudden change in the hereditary material, but *not* caused by segregation or recombination; the manner of origin of new hereditary variation.

nondisjunction The failure of chromosomes paired at meiosis to separate so that both members of one pair pass to the same gamete.

parthenogenesis Development of an egg without fertilization.

phenotype As defined by Johannsen, "the sum total of [an individual's] expressed characters." (*Cf. genotype.*)

polydactyly Having more than the normal number of fingers or toes.

proband method In human genetics, a method for comparing the proportion of children in families in which an "index case" or "proband" shows a specific trait or abnormality, with the proportion expected if the trait is inherited as a Mendelian unit.

pure line   The descendants, through self-fertilization, of a single homozygous parent; also, but less exactly, a line of plants or animals obtained by long-continued inbreeding. The essence of the pure-line concept is complete homozygosity, which is limited by the probability of new mutation.

reciprocal cross
   A ♀ × B ♂
   B ♂ × A ♀

root-cutting   Vegetative propagation of a plant by deriving a new individual from a piece of root.

self-incompatibility   (or *self-sterility*)   Failure of self-fertilization or of cross-fertilization between hermaphroditic plants or animals, usually due to inability of pollen (or sperm) to reach or penetrate an egg of an individual having a gene or genes in common with those of pollen.

self-sterility   (See *self-incompatibility*)

sex-linked   Referring to the transmission of a gene in the X chromosome, *e.g.*, in man or Drosophila—from mother to son.

sib method   In human genetics, the method of deriving the proportions of individuals with and without a trait under investigation from those members of a sibship *other* than the proband, or propositus, by which the sibship was brought to observation.

sport   Freak or mutation.

stirp   From Latin *stirpes* = root. The sum total of inherited elements; germ plasm.

synapsis   The pairing, at meiosis, of members of each homologous set of chromosomes, one having come from the mother, the other from the father.

syndactyly   Webbed fingers or toes.

xenia   The direct influence of foreign pollen on the endosperm of seed; *e.g.* in a white-kernelled variety of maize pollinated by a yellow-kernelled variety, the seeds arising from cross-pollination are yellow.

— Bibliography —

Original publications not seen are followed by an asterisk.

Avery, O. T., Colin M. MacLeod and Maclyn McCarty, 1944: "Studies on the Chemical Nature of the Substance Inducing Transformation of Pneumococcal Types," *J. Exp. Med.*, 79:137–158. (Reprinted in Peters, 1959.)

Babcock, E. B., 1949: "The Development of Fundamental Concepts in the Science of Genetics," *Portugaliae Acta Biologica*, Série A, Volume R. B. Goldschmidt: 1–46.

Baltzer, F., 1962: *Theodor Boveri, Leben und Werk eines Grossen Biologen*, Wissenschaftliche Verlagsgesellschaft, Stuttgart.

Barthelmess, Alfred, 1952: *Vererbungswissenschaft*, Orbis Academicus, Verlag Karl Alber, Freiburg/München.

Bateson, Beatrice, 1928: *William Bateson, F. R. S. Naturalist*, Cambridge University Press, London.

Bateson, William, 1894: *Materials for the Study of Variation*, Macmillan and Company, London.

———, 1902: *"Mendel's Principles of Heredity—a Defense,"* London (Reprinted in report of the Smithsonian Institution, Washington, D.C.)

———, 1907: "Facts Limiting the Theory of Heredity," (an address delivered at the International Zoological Congress, before the section of Cytology and Heredity, Aug. 23, 1907), *Science*, N.S., 26: 649–660.

———, 1908: *The Methods and Scope of Genetics, an Inaugural Lecture Delivered 23 October, 1908*, Cambridge University Press, London.

———, 1909: *Mendel's Principles of Heredity*, Cambridge University Press, London.

———, 1920: "Genetic Segregation," *Proc. Royal Soc., London*, 91: 358–368.

———, 1926: "Segregation: Being the Joseph Leidy Memorial Lecture of the University of Pennsylvania, 1922," *J. Genetics*, 16:201–235.

———, 1928: *The Scientific Papers of William Bateson*, R. C. Punnett, Ed., 2 vols., Cambridge University Press, London.

———, and R. C. Punnett, 1908: "The Heredity of Sex," *Science*, 27 (*cf.* Bateson 1928, pp. 179–182).

———, and E. R. Saunders, 1902: *Reports to the Evolution Committee of the Royal Society*, Report I (Dec. 17, 1901), Harrison & Sons, London.

————, E. R. Saunders, R. C. Punnett and C. C. Hurst, 1905: *Reports to the Evolution Committee of the Royal Society*, Report II, Harrison & Sons, London.

Baur, E., 1907: "Untersuchungen über die Erblichkeitsverhältnisse einer nur in Bastardform lebensfähigen Sippe von *Antirrhinum majus*," *Ber. deutsch. bot. Ges.*, 25:442–454.

————, 1927: "Die Experimentelle Erzeugung Leistungsfähigerer Rassen unserer Kulturpflanzen," *Die Naturwissenschaften*, 15:721–725.

Bauer, Hans, 1937: "Cytogenetik," *Fortschritte der Zoologie*, I:521–538.

Beadle, G. W., 1939: "Physiological Aspects of Genetics," *Ann. Rev. Physiol.*, 1:41–62.

————, 1959: "Genes and Chemical Reactions in Neurospora," *Science*, 129:1715–19.

————, 1963: *Genetics and Modern Biology, Memoirs Amer. Phil. Soc.*, 57, Philadelphia.

Belling, John, 1928: "The Ultimate Chromomeres of Lilium and Aloë with Regard to the Numbers of Genes," *Univ. of California Publications in Botany*, 14:307–318.

————, 1933: "Critical Notes on C. D. Darlington's *Recent Advances in Cytology*," *University of California Publications in Botany*, 17:75–110.

————, 1926: "Single and Double Rings at the Reduction Division in Uvularia," *Biol. Bull.*, 50:355–363.

Bernstein, F., 1924: "Ergebnisse einer biostatistischen zusammenfassenden Betrachtung über die erblichen Blutstrukturen des Menschen," *Klin. Wochenschr.*, 3:1495–1497.

Biffen, R. H., 1905: "Mendel's Law of Inheritance and Wheat Breeding," *Jour. Agr. Sci.*, 1:4–48.*

Blakeslee, Albert F., 1904: "Sexual Reproduction in the Mucorineae," *Proc. Amer. Acad. Arts and Sci.*, 40:205–319.

————, 1936: "Twenty-Five Years of Genetics (1910–1935)," *Brooklyn Bot. Gard. Memoirs*, 4:29–40.

————, 1941: "Chromosomal Interchanges," in *Cytology, Genetics and Evolution. University of Pennsylvania Bicentennial Conference*, Sept. 1940, University of Pennsylvania Press: 37–46.

Boveri, Theodor, 1902: "Über mehrpolige Mitosen als Mittel zur Analyse des Zellkerns," *Verh. der phys. med. Ges. Würzburg*, N F 35:67–90.

————, 1903: "Über die Konstitution der chromatischen Kernsubstanz." *Verh. deutsch. zool. Ges. 13 Vers. Würzburg* 10–33.

————, 1904: *Ergebnisse über die Konstitution der chromatischen Substanz des Zellkerns*, G. Fischer, Jena.

————, 1907: *Zellen-Studien VI. Die Entwicklung dispermer Seeigel Eier. Ein Beitrag zur Befruchtungslehre und zur Theorie des Kerns.* Tafel I-X, *Jenaische Zeits. für Naturwiss.*, 43:1–292.

Boyer, Samuel H., IV, 1963: *Papers on Human Genetics*, Prentice-Hall, Inc., Englewood Cliffs, New Jersey.

Bridges, C. B., 1913: "Non-disjunction of the Sex Chromosomes of Drosophila," *J. Exp. Zool*, 15:587–606.

————, 1916: "Non-disjunction as Proof of the Chromosome Theory of Heredity," *Genetics* 1:1–52; 107–163.

————, 1917: "Deficiency," *Genetics*, 2:445–465.

————, 1921: "Genetical and Cytological Proof of Non-disjunction of the Fourth Chromosome of Drosophila melanogaster," *Proc. Nat. Acad. Sci*, 7:186–192.

————, 1921: "Triploid Intersexes in Drosophila melanogaster," *Science*, 54:252–254.

————, 1922: "The Origin of Variations in Sexual and Sex-Limited Characters," *Amer. Nat.*, 56:51–63.

————, 1923: "The Translocation of a Section of Chromosome II upon Chromosome III in Drosophila," *Anat. Rec.*, 24:426–427.

————, 1935: "Salivary Chromosome Maps," *J. Hered.*, 26:60–64.

————, 1936: "The Bar 'Gene' a Duplication," *Science*, 83:210–211.

Carothers, E. Eleanor, 1913: "The Mendelian Ratio in Relation to Certain Orthopteran Chromosomes," *J. Morphol*, 24:487–511.

Caspari, Ernst, 1933: "Über die Wirkung eines pleiotropen Gens bei der Mehlmotte Ephestia Kühniella Zeller," *Wilhelm Roux' Arch. Entwick-Mech. der Org.*, 130:353–381.

Caspersson, T., 1936: "Über den chemischen Aufbau der Strukturen des Zellkernes," *Skand. Arch. Physiol.*, 73, Suppl. 8:1–151.

Castle, W. E., 1903a: "The Heredity of Sex," *Bull. Mus. Comp. Zool, Harvard College*, 40:189–218.

————, 1903b: "The Laws of Heredity of Galton and Mendel, and some Laws governing Race Improvement by Selection," *Proc. Amer. Acad. Arts and Sci.*, 39:223–242.

————, 1905: "Heredity of Coat Characters in Guinea Pigs and Rabbits," *Carnegie Inst. Washington Publ.* 23:1–78.

————, 1951: "The beginnings of Mendelism in America," in L. C. Dunn, Ed., *Genetics in the 20th Century*, The Macmillan Company, New York.

————, and C. C. Little, 1910: "On a Modified Mendelian Ratio among Yellow Mice," *Science*, 32:868–870.

————, and J. C. Phillips, 1909: "A Successful Ovarian Transplantation in the Guinea-Pig, and its Bearing on Problems of Genetics," *Science*, 30:312–313.

Chetverikov, S. S., 1926: "On Certain Aspects of the Evolutionary Process from the Standpoint of Modern Genetics" (in Russian), *Zh. Eks. Biol.*, A2:3–54. (English translation edited with note by I. M. Lerner in *Proc. Amer. Phil. Soc.*, 105:167–195 (1961).)

————, 1927: "Über die genetische Beschaffenheit wilder Populationen," *Verh. 5te Kong. Vererb. Berlin*, 2:1499–1500.

Coleman, William, 1965: "Cell, Nucleus and Inheritance: an Historical Study," *Proc. Amer. Philos. Soc.*, 109:124–158.

Correns, Carl, 1899: "Untersuchungen über die Xenien bei Zea mays." *Ber. deutsch. bot. Ges.*, 17:410–417.

————, 1900a: "G. Mendels Regel über das Verhalten der Nachkommenschaft der Rassenbastarde," *Ber. deutsch. bot. Ges.,* 18:158–168. (English translation in *Genetics,* 35 (1950), Suppl. 33–41.

———, 1900b: "Über Levkojenbastarde: Zur Kenntniss der Grenzen der Mendelschen Regeln," *Botanisches Centralblatt,* 84:97–113.

———, 1902a: "Scheinbare Ausnahmen von der Mendelschen Spaltungsregel für Bastarde," *Ber. deutsch. bot. Ges.,* 20:159–172.

———, 1902b: "Über den Modus und den Zeitpunkt der Spaltung der Anlagen bei den Bastarden vom Erbsen-Typus." *Bot. Zeit. Jg.,* 60 Abt II:65–82.

———, 1905: "Gregor Mendels Briefe an Carl Nägeli 1866–1873. Ein Nachtrag zu den veröffentlichten Bastardierungsversuchen Mendels," *Abh. d. Math-Phy. Kl.d.Konigl. Sachs. Ges. d. Wiss,* 29 Nr.3:189–265 (English translation in *Genetics,* 35 (1950), suppl. to No. 5, Part 2:1–29.)

———, 1907: *Die Bestimmung und Vererbung des Geschlechtes,* Gebrüder Borntraeger, Berlin.

———, 1921: "Die ersten zwanzig Jahre Mendelscher Vererbungslehre," *Festschr. der Kaiser Wilhelm Gesellschaft:* 42–49.

———, 1922: "Etwas über Gregor Mendels Leben und Wirken, *in* Dem Andenken an Gregor Mendel zur Jahrhundertfeier seines Geburstages," *Die Naturwiss.,* 29:621–631.

———, 1924: *Gesammelte Abhandlungen zur Vererbungswissenschaft aus periodischen Schriften,* F. von Wettstein, Ed., J. Springer, Berlin. (All of Correns's papers referred to in the text will be found here except the 1937 paper.)

———, 1937: "Nichtmendelnde Vererbung," *Handbuch der Vererb. Wiss.,* 2 H Gebrüder Borntraeger, Berlin.

Creighton, H. B., and B. McClintock, 1931: "A correlation of Cytological and Genetical Crossing Over in *Zea mays*," *Proc. Nat. Acad. Sci.,* 17:492–497. (Reprinted in Peters, 1959.)

———, and Barbara McClintock, 1932: "Cytological Evidence for 4-Strand Crossing Over in *Zea mays*," *Proc. 6th Int. Congr. Genetics,* 2:392. (See also Rhoades, M. M. and Barbara McClintock, 1935.)

Crowther, J. G., 1952: *British Scientists of the 20th Century,* Routledge & Kegan Paul, Ltd., London. Life of Wm. Bateson, pp. 248–310.

Cuénot, Lucien, 1902: "La loi de Mendel et l'hérédité de la pigmentation chez les souris," *Arch. zool. exp. et gén,* 3° ser. 10:Notes XXVII–XXX.

———, 1903: "L'hérédité de la pigmentation chez les souris," 2ᵐᵉ Note *Arch. zool. exp. et gén.,* 4° ser., Notes XXXIII–XXXVIII.

———, 1905: "Les races pures et les combinaisons chez les souris," *Arch. zool. exp. et gén.,* 4 Série III, Notes CXXIII–CXXXII.

Danforth, C. H., 1923: "The Frequency of Mutation and the Incidence of Hereditary Traits in Man," *Eugenics, Genetics and the Family. Second International Congress on Eugenics,* 1:120–128, New York, 1921.

Darbishire, A. D., 1904: "On the Result of Crossing Japanese Waltzing with Albino Mice," *Biometrika*, 3:1–51.

Darlington, C. D., 1937: *Recent Advances in Cytology*. 2nd ed., Churchill, London.

_____, 1939: *Evolution of Genetic Systems*, Cambridge University Press, London.

_____, 1960: "Chromosomes and the Theory of Heredity," *Nature*, 187:892–895. (Reprinted in Smithsonian Report for 1961:417–427.)

Darwin, Charles, 1876 (1868): *The Variation of Animals and Plants Under Domestication*, 2nd ed., revised, 2 vols., D. Appleton and Company, New York.

Darwin, Francis, 1887: *Life and Letters of Charles Darwin*, 2 vols. D. Appleton and Company, New York.

Davenport, C. B., 1901: "Mendel's Law of Dichotomy," *Biological Bulletin*, 2:307.

Dobzhansky, Th., 1929a: "Genetical and Cytological Proof of Translocations involving the Third and Fourth Chromosomes of *Drosophila melanogaster*," *Biol. Zbl.*, 49:408–419.

_____, 1929b: "A Homozygous Translocation in *Drosophila melanogaster*," *P.N.A.S.*, 15:633–638.

_____, 1936: "Position Effects on Genes," *Biol. Rev.*, 11:364–382.

_____, 1937: *Genetics and the Origin of Species*, Columbia University Press, New York.

Dodge, B. O., 1927: "Nuclear Phenomena Associated with Heterothallism and Homothallism in the Ascomycete Neurospora," *J. Agr. Res.*, 35:289–305.

Dodson, Edward O., 1955: "Mendel and the Rediscovery of his Work," *Scientific Monthly*, 81:187–195.

Doncaster, L., 1908: "On Sex-Inheritance in the Moth *Abraxas grossulariata* and its Variety Lacticolor," *Roy. Soc. Rpts. to the Evolution Committee*, IV:53–57.

Dorsey, M. J., 1944: "Appearance of Mendel's Paper in American Libraries," *Science*, 99:199–200.

Driesch, Hans, 1928: *Philosophie des Organischen*, 4th ed., Berlin.

Dunn, L. C., 1917: "Nucleus and Cytoplasm as Vehicles of Heredity," *Amer. Nat.*, 51:286–300.

_____, 1921: "Unit Character Variation in Rodents," *J. Mammal*, 2:125–140. (Reprinted in Peters, 1959.)

_____, 1962: "Cross Currents in the History of Human Genetics," *Amer. J. Hum. Genetics*, 14:1–13.

_____, 1964: "Old and New in Genetics," *Bull. N.Y. Acad. Med.*, 40:325–333.

_____, 1965a: "Biographical Memoir of William Ernest Castle 1867–1962," *Nat. Acad. Sci. Biog. Mem.*, 45:33–80.

_____, 1965b: "Mendel, His Work and His Place in History," *Proc. Amer. Philos. Soc.*, 109: (in press).

East, Edward M., 1907: "The Relation of Certain Biological Principles to Plant Breeding," *Conn. Agr. Exp. Sta. Bull.*, 158:1–93.

————, 1910: "A Mendelian Interpretation of Variation that is Apparently Continuous," *Amer. Nat.*, 44:65–82.

————, 1918: "The Role of Reproduction in Evolution," *Amer. Nat.*, 52:273–289.

————, and H. K. Hayes, 1911: "Inheritance in Maize," *Conn. Agr. Exp. Sta. Bull.*, 167:1–141.

————, and H. K. Hayes, 1912: "Heterozygosis in Evolution and in Plant Breeding," *U.S. Dept. Agr. Bur. Plant Ind. Bull.*, 243:1–58.

————, and D. F. Jones, 1919: *Inbreeding and Outbreeding*, Lippincott, Philadelphia.

Edwardson, J. R., 1962: "Another Reference to Mendel Before 1900," *J. Hered.*, 53:152.

Eichling, C. W., 1942: "I Talked with Mendel," *J. Hered.*, 33:243–246.

Emerson, R. A., G. W. Beadle and A. C. Fraser, 1935: "A Summary of Linkage Studies in Maize," *Cornell Univ. Agr. Exp. Sta. Mem.*, 80:1–83.

Emerson, S., 1935: "The Genetic Nature of de Vries' Mutations in *Oenothera lamarckiana*," *Amer. Nat.*, 69:545–559.

Farabee, W. C., 1905: "Inheritance of Digital Malformations in Man," *Papers, Peabody Museum of Harvard University*, 3:69–77, Cambridge, Mass.

Fish, Harold D., 1914: "On the Progressive Increase of Homozygosis in Brother-Sister Matings," *Amer. Nat.*, 48:759–761.

Fisher, R. A., 1918: "The Correlation Between Relatives on the Supposition of Mendelian Inheritance," *Roy. Soc. Edin. Trans.*, 52:399–433.

————, 1922: "On the Dominance Ratio." *Proc. Roy. Soc. Edin.*, 52:321–341.

————, 1930: *The Genetical Theory of Natural Selection*, Clarendon Press, Oxford.

————, 1932: "The Bearing of Genetics on Theories of Evolution," *Science Progress*, 106:1–16.

————, 1936: "Has Mendel's Work been Rediscovered?" *Annals of Science*, 1:115–137.

————, and E. B. Ford, 1950: "The 'Sewall Wright Effect'," *Heredity*, 4:117–119.

Focke, Wilhelm Olbers, 1881: *Die Pflanzenmischlinge, eine Beitrag zur Biologie der gewächse*," Berlin.

Galton, Francis, 1869: *Hereditary Genius*, London.

————, 1871: "Experiments in Pangenesis," *Proc. Roy. Soc.*, 19:393.

————, 1876: "A Theory of Heredity," *J. Anthropol. Inst.*, 5:329–348.

————, 1889: *Natural Inheritance*, The Macmillan Company, New York.

Garrod, Archibald E., 1899: "A Contribution to the Study of Alkaptonuria," *Med. Chir. Trans.*, 82:369–394.

————, 1901: "About Alkaptonuria," *Lancet*; ii:1484–1486.

————, 1902: "The Incidence of Alkaptonuria: a Study in Chemical Individuality," *Lancet*, ii:1616–1620. (Reprinted in Harris, 1963).

————, 1908: "Croonian Lectures to the Royal Academy of Medicine: Inborn Errors of Metabolism," *Lancet* ii:1–7; 142–148; 173–179; 214–220.

————, 1909: *Inborn Errors of Metabolism* (a revision of the Croonian Lectures of 1908), Frowde and Hodder and Stoughton, London.

Gasking, E. B., 1959: "Why Was Mendel's Work Ignored?" *J. Hist. Ideas*, **20**:60–84.

Gaupp, E., 1917: *August Weismann sein Leben und sein Werk*, G. Fischer, Jena.

Gerould, John H., 1921: "Blue-Green Caterpillars," *J. Exp. Zool.*, **34**:385–415.

Glass, H. B., 1947: "Maupertuis and the Beginning of Genetics," *Quart. Rev. Biol.*, **22**:196–210.

————, 1953: "The Long Neglect of a Scientific Discovery: Mendel's Laws of Inheritance," in *Studies in Intellectual History*, G. Boas, Ed., Johns Hopkins Press, Baltimore, 148–160.

————, 1963: "The Establishment of Modern Genetical Theory as an Example of the Interaction of Different Models, Techniques and Inferences," in *Scientific Change*, A. C. Crombie, Ed., Basic Books, New York.

————, 1957: "Review of Barthelmess' *Vererbungswissenschaft*," *Quart, Rev. Biol.*, **32**:376.

Goldschmidt, Richard, 1927: *Physiologische Theorie der Vererbung*, Springer, Berlin.

————, 1934: "Lymantria," *Bibliogr. Genet.*, **11**:1–186.

————, 1938: *Physiological Genetics*, McGraw-Hill Book Company, Inc., New York.

Goodale, H. D., 1909: "Sex and its Relation to the Barring Factor in Poultry," *Science*, **29**:1004–1005.

Griffith, F., 1928: "The Significance of Pneumococcal Types," *J. Hygiene*, **27**:113–159.

Gulick, J. T., 1872: "On Diversity of Evolution under one Set of External Conditions," *J. Linnean Soc., London, Zool.*, **11**:496–505.

————, 1888: "Divergent Evolution Through Cumulative Segregation," *J. Linnean Soc., London, Zool.*, **20**:189–274. (Reprinted in *Reports of Smithsonian Inst.*, 1891:269–336.)

————, 1905: "Evolution, Racial and Habitudinal," *Carnegie Inst. Washington Publ.*, **25**:1–269.

Gunther, M., and L. S. Penrose, 1935: "The Genetics of Epiloia," *J. Genet.*, **31**:413–430.

Gustafsson, A., 1963: "Mutations and the Concept of Viability," In *Recent Plant Breeding Research*, John Wiley and Sons, New York.

Haacke, W., 1906: "Die Gesetze der Rassenmischung und die Konstitution des Keimplasmas. W. Roux's *Arch. f. Entwick—Mech. d. Org.* **21**:1–93.

Haecker, Valentin, 1918: *Entwicklungsgeschictliche Eigenschaftsanalyse (Phänogenetik)*, G. Fischer, Jena.

Haldane, J. B. S., 1928: *Possible Worlds & other papers*, Harper and Bros., New York & London.

———, 1932: *The Causes of Evolution*, Longmans Green, London.

———, 1930–1932: "A Mathematical Theory of Natural and Artificial Selection," *Proc. Cambridge Philosophical Society*, **26–28**, Parts I–IX.

———, 1935: "The Rate of Spontaneous Mutation of a Human Gene," *J. Genet.*, **31**:317–326.

———, 1938: "Forty years of Genetics," in Needham, J., and W. Pagel, editors, *Background to Modern Science*, Cambridge (England) and New York.

———, 1954: *The Biochemistry of Genetics*, George Allen and Unwin, Ltd., London.

———, 1958: "Karl Pearson 1857–1954," *Biometrika*, **44**:303–313.

Hämmerling, J., 1934: "Entwicklungsphysiologische und genetische Grundlagen der Formbildung bei der Schirmalge Acetabularia," *Die Naturwissenschaften*, **22**:829–836.

Hardy, G. H., 1908: "Mendelian Proportions in a Mixed Population," *Science*, **28**:49–50 (reprinted in Peters, 1959).

Harris, H., 1963: *Garrod's "Inborn Errors of Metabolism,"* Oxford University Press (also contains Garrod, 1902 and 1909).

Heimans, J., 1962: "Hugo de Vries and the Gene Concept," *Amer. Nat.*, **96**:93–104.

Hervé, G., 1912: "Maupertuis Généticien," *Rev. Anthropol.*, **22**:217–230.

Hirszfeld, L., and Hanka Hirszfeld, 1919: "Serological Differences between the Blood of Different Races. The Result of Researches on the Macedonian Front," *Lancet*, **ii**:675–679.

Hirszfeldowa, H., A. Kelus and F. Milgrom, 1956: "Ludwik Hirszfeld," *Travaux Soc. Sci. et Lettres de Wroclaw*.

Hoffmann, Hermann, 1869: "Untersuchungen zur Bestimmung des Werthes von Species und Varietät: ein Beitrag zur Kritik der Darwin'-schen Hypothese," Giessen.*

Hoge, M. A., 1915: "Another Gene in the Fourth Chromosome of Drosophila," *Amer. Nat.*, **49**:47–49.

Hughes, A., 1959: *A History of Cytology*, Abelard-Schuman, London.

Iltis, Hugo, 1924: *Gregor Mendel: Leben, Werk and Wirkung*. Springer, Berlin. (English translation by Eden and Cedar Paul: *Life of Mendel*, W. W. Norton Company, New York, 1932.)

Ingram, V. M., 1957: "Gene Mutations in Human Haemoglobin: the Chemical Difference between Normal and Sickle-Cell Haemoglobin," *Nature*, **180**:326–328 (reprinted in Boyer, 1963).

Janssens, F. A., 1909: "La Théorie de la chiasmatypie, nouvelle interpretation des cinéses de maturation," *La Cellule*, **25**:387–406.

Jenkin, Henry Charles Fleeming, 1867: "Origin of Species," *North British Review*, **46**:277–318. This review article is untitled and unsigned but is identified by the reference to. it in F. Darwin, 1887, Vol. II, p. 288.

Jennings, H. S., 1914: "Formulae for the Results of Inbreeding," *Amer. Nat.*, **48**:693–696.

———, 1916: "The Numerical Results of Diverse Systems of Breeding," *Genetics*, 1:53–89.

———, 1917: "The Numerical Results of Diverse Systems of Breeding with Respect to Two Pairs of Characters," *Genetics*, 2:97–154.

———, 1923: "The Numerical Relations in the Crossing Over of the Genes, with a Critical Examination of the Theory that the Genes Are Arranged in a Linear Series," *Genetics*, 8:393–457.

———, 1925: *Prometheus, or Biology and the Advancement of Man*, E. P. Dutton and Co., New York.

Johannsen, W., 1896: *Om Arvelighed og Variabilitet*, Det Schubotheske Forlag, København.

———, 1903: *Om Arvelighed i Samfund og i rene Linier. Oversigt over det Kgl. danske videnskabernes selskabs forhandlinger #3 forelagt i modet den 6 Feb. 1903*. (English translation in Peters, 1959.)

———, 1905: "Arvelighedslaerens Elementer; forelaesninger holdte ved Københavns Universitet," *Gyldendalske Boghandel*, Nordiske Forlag, København og Kristiania.

———, 1909: *Elemente der Exakten Erblichkeitslehre*, G. Fischer, Jena.

———, 1915: "Experimentelle Grundlagen der Deszendenzlehre: Variabilität, Vererbung, Kreuzung, Mutation," in *Kultur der Gegenwart III, IV 1, Allgemeine Biologie*: 597–661.

———, 1922: "Biologi: Traek af de biologiske Videnskabers Udvikling i det nittende Aarhundrede," *Gyldendalske Boghandel*, Nordisk Forlag, København.

———, 1923: "Hundert Jahre Vererbungsforschung," *Verh. Ges. deutscher Naturf. und Ärtzte*, 87, Versamml. 70–104.

Jones, D. F., 1959: "Basic Research in Plant and Animal Improvement," *Proc. 10th Int. Congr. Genetics*, I:172–176.

Koltzoff, N. K., 1939: "Les molécules héréditaires," *Actualités scientifiques et industrielles*, 776:1–60. Herman et Cie, Editeurs, Paris.

Kornberg, A., 1962: *Enzymatic Synthesis of DNA*, John Wiley and Sons, New York.

Krizenecky, Jaroslaw, 1963: "Mendels zweite erfolglose Lehramtsprüfung in Jahre 1856," *Sudhoff Arch. für Geschichte der Medizin und Naturwiss.*, 47:305–310.

Krumbiegel, Ingo, 1957: *Gregor Mendel und das Schicksal seiner Entdeckung*, Wissenschaftliche Verlagsgesellschaft M.B.H. (Band 22 in *Grosse Naturforscher*), Stuttgart.

Landsteiner, Karl, 1901: "Über Agglutinationserscheinungen normalen menschlichen Blutes," *Wiener Klinische Wochenschrift*, 14: 1132–1134. (Reprinted in Boyer, 1964.)

Lang, Arnold, 1914: *Die Experimentelle Vererbungslehre in der Zoologie seit 1900*, G. Fischer, Jena.

Lanni, Frank, 1964: "The Biological Coding Problem," *Adv. in Genetics*, 12:1–141.

Lawrence, W. J. C. and J. R. Price, 1940: "The Genetics and Chemistry of Flower Colour Variation," *Cambridge Philos. Soc. Biol. Rev.*, 15:35–58.

Lehmann, Ernst, 1916: "Aus der Frühzeit der pflanzlichem Bastardierungskunde," *Arch. Gesch. d. Naturwiss. u.d. Technik,* p. 78.

Lock, R. H., 1906: *Recent Progress in the Study of Variation, Heredity and Evolution,* John Murray, London.

McClung, Clarence E., 1902: "The Accessory Chromosome—Sex Determinant?" *Biol. Bull.,* 3:43–84.

McKusick, V. A., 1960: "Walter S. Sutton and the Physical Basis of Mendelism," *Bull. Hist. Med.,* 34:487–497.

Mangelsdorf, P. C., 1951: "Hybrid Corn: Its Genetic Basis and Its Significance in Human Affairs," in L. C. Dunn, Ed., *Genetics in the 20th Century,* The Macmillan Company, New York.

Mayer, C. F., 1953: "Genesis of Genetics," *Acta Genet. Med. et Gemello.,* 2:237–332.

Mayr, Ernst, 1963: *Animal Species and Evolution,* Harvard University Press, Cambridge, Massachusetts.

Mendel, Gregor, 1866: "Versuche über Pflanzenhybriden," *Verh. naturforsch. Verein Brünn,* 4:3–47. (Reprinted in same journal 1911, Vol. 49; in *Flora,* 89 (1901); and in Ostwald's *Klassiker der exakten Wissenschaften,* No. 121, 1900. (English translation in Bateson, 1909 and in Sinnott, Dunn and Dobzhansky, 5th ed., 1958.)

———, 1870: "Über einige aus künstlicher Befruchtung gewonnene Hieracium Bastarde," *Verh. naturforsch. Verein Brünn,* 8:26–31. English translation in Bateson, 1909.

———, 1871: "Die Windhose vom 13. October, 1870,"*Verh. naturforsch. Verein Brünn,* 9:229–246.

Merton, Robert K., 1961: "Singletons and Multiples in Scientific Discovery: A Chapter in the Sociology of Science," *Proc. Amer. Philos. Soc.,* 105:470–486.

Merz, John T., 1903: *A History of European Thought in the 19th Century,* Vol. II, *Scientific Thought,* Blackwood, London and Edinburgh.

Montgomery, T. H., Jr., 1901: "A Study of the Chromosomes of the Germ-Cells of Metazoa," *Trans. Amer. Phil. Soc.,* 20 (as cited by Sutton, 1903).*

Moore, J. A., 1963: *Heredity and Development,* Oxford University Press, New York.

Morgan, T. H., 1905: "The Assumed Purity of the Germ Cells in Mendelian Results," *Science,* N. S., 22:877–879.

———, 1910a: "Sex-Limited Inheritance in Drosophila," *Science,* 32: 120–122 (reprinted in Peters, 1959, pp. 63–66).

———, 1910b: "The Method of Inheritance of Two Sex-Limited Characters in the Same Animal," *Proc. Soc. Exp. Biol. Med.,* 8:17–19.

———, 1910c: "Chromosomes and Heredity," *Amer. Nat.,* 44:449–496.

———, 1911: "An Attempt to Analyse the Constitution of the Chromosomes on the Basis of Sex Limited Inheritance in Drosophila, *J. Exp. Zool.,* 11:365–413.

———, 1914: *Heredity and Sex, the Jesup Lectures 1913,* 2nd (revised) edition, Columbia University Press, New York.

———, A. H. Sturtevant, H. J. Muller and C. B. Bridges, 1915: *The Mechanism of Mendelian Heredity,* Henry Holt and Co., New York.

———, 1923: "The Modern Theory of Genetics and the Problem of Embryonic Development," *Phys. Rev.,* 3:603–627.

———, 1926: *The Theory of the Gene,* Yale University Press, New Haven.

———, 1932: "The Rise of Genetics," *Science,* 76:261–267 and 285–288.

———, 1934: *Embryology and Genetics,* Princeton University Press, Princeton.

———, 1939: "Personal Recollections of Calvin B. Bridges," *J. Hered.,* 30:354–358.

———, 1940: "Biographical Memoir of Edmund Beecher Wilson 1856–1939," *Nat. Acad. Sci. Biog. Mem.,* 21:315–342.

———, 1941: "Biographical Memoir of Calvin Blackman Bridges 1889–1938," *Nat. Acad. Sci. Biog. Mem.,* 22:31–48.

Motulsky, A. G., 1958: "Josef Adams (1756–1818), the Forgotten Founder of Medical Genetics," *Proc. X Int. Cong. Genet.,* II:198.

Muller, H. J., 1916: "The Mechanism of Crossing Over," *Amer. Nat.,* 50:193–221; 284–305; 350–366; 421–434.

———, 1918: "Genetic Variability, Twin Hybrids and Constant Hybrids in a Case of Balanced Lethal Factors," *Genetics,* 3:422–499.

———, 1922: "Variation Due to Change in the Individual Gene," *Amer. Nat.,* 56:32–50. (Reprinted in Peters, 1959.)

———, 1927: "Artificial Transmutation of the Gene," *Science,* 66:84–87.

———, 1943: "Edmund B. Wilson—an Appreciation," *Amer. Nat.,* 77:5–37 and 142–172.

———, 1947: "The Production of Mutations," *J. Hered.,* 38:259–270 (Nobel lecture).

———, and T. S. Painter, 1929: "The Cytological Expression of Changes in Gene Alignment Produced by X-rays in Drosophila," *Amer. Nat.,* 63:193–200.

Nachtsheim, Hans, 1960: "Die Entwicklung der Genetik in Deutschland von der Jahrhundertwende bis zum Atomzeitalter," *Studium berolinense,* 1960:858–867.

Naegeli, Carl von, 1884: *Mechanisch-physiologische Theorie der Abstammungslehre,* R. Oldenburg, München u. Leipzig.

Needham, Joseph, 1962: "Frederick Gowland Hopkins," *Perspectives in Biol. and Med.,* 6:2–46.

Nilsson-Ehle, H., 1909: "Kreuzungsuntersuchungen an Hafer und Weizen," *Lunds Univ. Aarskr. N. F. Afd.,* 2,5,2:122 pp.*

Painter, T. S., 1933: "A New Method for the Study of Chromosome Rearrangements and the Plotting of Chromosome Maps," *Science,* 78:585–586.

———, and H. J. Muller, 1929: "Parallel Cytology and Genetics of Induced Translocations and Deletions in Drosophila," *J. Hered.*, 20:287–298.

Pauling, Linus, Harvey A. Itano, S. J. Singer and Ibert C. Wells, 1949: "Sickle Cell Anemia, a Molecular Disease," *Science*, 110: 543–548 (reprinted in Boyer, 1963).

Pearl, Raymond, 1911: "Biometric Ideas and Methods in Biology," *Scientia*, 10:101–119.

———, 1913: "A Contribution towards an Analysis of the Problem of Inbreeding," *Amer. Nat.*, 47:577–614.

———, 1915: *Modes of Research in Genetics*, The Macmillan Company, New York.

Pearson, Karl, 1904: "On a Generalized Theory of Alternative Inheritance, with Special Reference to Mendel's Laws." *Phil. Trans. Roy. Soc. London*, A203:53–86.

———, 1914–1930: *The Life, Letters and Labors of Francis Galton*, Vols. I, II, III, IIIa, Cambridge University Press, London.

Penrose, L. S., 1932: "On the Interaction of Heredity and Environment in the Study of Human Genetics (with special reference to Mongolian imbecility)," *J. Genetics*, 25:407–422.

Peters, James A., 1959: *Classic Papers in Genetics*, Prentice-Hall, Inc., Englewood Cliffs, N.J.

Plough, Harold H., 1917: "The Effect of Temperature on Crossing Over in Drosophila," *J. Exp. Zool.*, 24:147–208.

Punnett, R. C., 1911: *Mendelism*, 3rd ed., The Macmillan Company, New York.

———, 1915: *Mimicry in Butterflies*, Cambridge University Press, London.

———, 1950: "Early Days of Genetics," *Heredity*, 4:1–10 (Apr.). (An address delivered at the 100th meeting of the Genetical Society, June 30, 1949.)

Renner Otto, 1959: *"Botanik"* in *Geist u. Gestalt*, 2. Band *Naturwiss. Biograph. Beitr. z. Gesch. d. Bayer. Adak. d. Wiss. vornehmlich in 2. Jahrhundert ihres Bestehens.*, C. H. Becksche Verlagsbuchhandlung, München.*

———, 1961: "William Bateson und Carl Correns," *Sitzber. Heidelberg. Akad. Wiss. Math-Natur-Wiss. Kl.* 60/61, 6:159–184.

Rhoades, M. M. and Barbara McClintock, 1935: "The Cytogenetics of Maize," *Bot. Rev.*, 1:292–325.

Richter, Oswald, 1925: "Biographisches über Pater Gregor Mendel aus Brünns Archiven, in Ruzicka, V., Ed., *Memorial Volume in Honor of the 100th Birthday of J. G. Mendel*, Fr. Borovy, Prague.

———, 1932: Gregor Mendel's Reisen. *Verh. Naturf. Verein Brünn* 63:1–10.

———, 1940: "75 Jahre seit Mendel's Grosztat und Mendels Stellungnahme zu Darwins werken auf Grund Seiner Entdeckungen," *Verh. naturf. Ver. Brünn.*, 72:110–173.

———, 1943: "Johann Gregor Mendel wie er wirklich war. Neue Beiträge zur Biographie des berühmten Biologen aus Brünns Archiven. Herausgegeben mit Unterstützung des mährischen Landesbehörde, der Landeshauptstadt Brünn und der Deutschen Akademie der Wissenschaften in Prag. Druck von Josef Klar, Brünn," *Verh. naturf. Verein Brünn,* **74.**

Roberts, H. F., 1929: *"Plant Hybridization Before Mendel,"* Princeton University Press, Princeton.

Rostand, J., 1953: *Instruire sur l'homme,* La Diane Francaise, Paris.

———, 1956: "Esquisse d'une histoire de l'atomisme en biologie," Ch. 1 in *L'atomisme en biologie,* Gallimard, Paris.

Rudorf, W., Ed., 1959: *Dreissig Jahre Züchtungsforschung. Zum Gedenken an Erwin Baur (16.4.1874–2.12.1933),* G. Fischer Verlag, Stuttgart.

Russell, E. S., 1930: *The Interpretation of Development and Heredity,* Clarendon Press, Oxford.

Sageret, Augustin, 1826: "Considerations sur la production des hybrids des variantes et des varietés en general, et sur celles des Cucurbitaceés en particulier," *Annales des Sci. Nat. Prem. Série,* **8:** 294–314.*

Sajner, J., 1963: "Gregor Mendels Krankheit und Tod," *Sudhoffs Arch. für Geschichte der Medizin und Naturwiss.,* **47:**372–382.

Schleip, Waldemar, 1927: *Entwicklungsmechanik und Vererbung bei Tieren,* Gebrüder Boentraeger, Berlin (in *Handbuch der Vererbungswissenschaft.*

Schultz, J. and Caspersson, T., 1939: "Heterochromatic Regions and the Nucleic Acid Metabolism of the Chromosomes," *Arch. exp. Zellforsch.,* **22:**650–654.

Shull, G. H., 1908: "The Composition of a Field of Maize," *Report Amer. Breeders' Ass.,* **4:**296–301.

———, 1909: "A Pure Line Method of Corn Breeding," *Report Amer. Breeders' Ass.,* **5:**51–59.

———, 1911: "Reversible Sex Mutants in *Lychnis dioica,*" *Bot. Gaz.,* **52:**329–368.

———, 1952a: "Beginnings of the Heterosis Concept," in Gowen, J. W., Ed., *Heterosis,* Iowa State College Press, Ames, Iowa.

———, 1952b: *Erich von Tschermak-Seysenegg* (biographical note), *Genetics,* **37:**1–7.

Sinnott, E. W., L. C. Dunn and Th. Dobzhansky, 1958: *Principles of Genetics,* 5th ed., McGraw-Hill Book Company, Inc., New York.

Sirks, M. J., 1952: "The Earliest Illustrations of Chromosomes," *Genetica,* **26:**65–76.

———, 1956: *General Genetics,* English translation of the 5th Dutch edition by J. Weijer and D. Weijer-Tolmee, M. Nijhoff, The Hague.

Slizynska, H., 1938: "Salivary Chromosome Analysis of the White-Facet Region of Drosophila melanogaster," *Genetics,* **22:**291–299.

Snyder, L. H., 1959: "Fifty Years of Medical Genetics," *Science,* **129:**7–13.

Spencer, Herbert, 1864: *Principles of Biology,* London.

Spillman, W. J., 1901: "Quantitative Studies on the Transmission of Parental Characters of Hybrid Offspring," *Proc. 15th Ann. Con-*

vention Assn. of Amer. Agr. Colleges etc., Nov. 12–14, 1901, U. S. Dept. Agric. Office of Exp't. Stations Bulletin, **115**:88–98.

————, 1909: "Barring in Barred Plymouth Rocks," *Poultry*, **5**:708 (August).

————, 1910: "Mendelian Phenomena without de Vriesian Theory," *Amer. Nat.*, **44**:214–228.

————, 1911: "Application of Some of the Principles of Heredity to Plant Breeding," *U. S. Dept. Agric. Bureau of Plant Industry Bulletin*, **195**.

Sprengel, Christian Konrad, 1793: *Das entdeckte Geheimnis der Natur im Bau und in der Befrüchtung der Blumen* (cf. Roberts, 1929).*

Stadler, L. J., 1928: "Mutations in Barley Induced by X-Rays and Radium," *Science*, **68**:186–187.

————, 1954: "The Gene," *Science*, **120**:811–819.

Stein, Emmy, 1950: "Dem Gedächtnis von Carl Erich Correns nach einem halben Jahrhundert der Vererbungswissenschaft," *Die Naturwissenschaften*, **37**:457–463.

Stern, Curt, 1931: "Zytologisch-genetische Untersuchungen als Beweise für die Morgansche Theorie des Factorenaustauschs," *Biol. Zbl.*, **51**:547–587.

————, 1950: "Boveri and the Early Days of Genetics," *Nature*, **166**:446.

————, 1962: "Wilhelm Weinberg 1862–1937," *Genetics*, **47**:1–5. (A bibliography of Weinberg's writings 1886–1935 to accompany this article was prepared by Eva R. Sherwood and is available in mimeographed copies from the Dept. of Zoology, University of California, Berkeley.)

Stomps, Th. J., 1954: "On the Rediscovery of Mendel's Work by Hugo de Vries," *J. Hered.*, **45**:293–294.

Stubbe, Hans, 1938: *Genmutation: I, allgemeiner Teil, Handbuch der Vererbwiss.*, Bd. IIF:1–429, Gebrüder Borntraeger, Berlin.

————, 1963: *Kurze Geschichte der Genetik bis zur Wiederentdeckung der Vererbungsregeln Gregor Mendels*, G. Fischer, Jena.

Sturtevant, A. H., 1913: "The Himalayan Rabbit Case, with some Considerations on Multiple Allelomorphs," *Amer. Nat.*, **47**:234–239.

————, 1920: "The Vermilion Gene and Gynandromorphism," *Proc. Soc. Exp. Biol. Med.*, **27**:70–71.

————, 1925: "The Effects of Unequal Crossing Over at the Bar Locus in Drosophila," *Genetics*, **10**:117–147.

————, 1959: "Thomas Hunt Morgan," *Nat. Acad. Sci. Biog. Mem.*, **33**:282–325.

————, and G. W. Beadle, 1936: "The relations of inversions in the X-chromosome of *Drosophila melanogaster* to crossing over and disjunction," *Genetics*, **21**:554–604.

————, and G. W. Beadle, 1939: *An Introduction to Genetics*, W. B. Saunders, Philadelphia. (Reprinted by Dover Books, New York, 1962.)

————, and T. H. Morgan, 1923: "Reverse Mutation of the Bar Gene Correlated with Crossing Over," *Science*, **57**:746–747.

Sumner, F. B., 1932: "Genetic, Distributional and Evolutionary Studies of the Subspecies of Deer-mice (Peromyscus)," *Bibliog. Genet.*, **9**:1–106.

Sutton, W. S., 1902: "On the Morphology of the Chromosome Group of *Brachystola magna*," *Biol. Bull.*, **4**:24–39.

————, 1903: "The Chromosomes in Heredity," *Biol. Bull.*, **4**:231–251.

Timoféeff-Ressovsky, N. W., K. G. Zimmer and M. Delbrück, 1935: "Über die Natur der Genmutation und der Genstruktur," *Nachr. Ges. Wiss. Göttingen, Math.-Phys. Kl. 6 Biol.*:189–245.

————, 1937: *Experimentelle Mutationsforschung in der Vererbungslehre*, Steinkopff, Dresden und Leipzig.

————, 1938: "Genetica di popolazioni," *La Ricerca Scientifica*, Serie II Anno IX **1**:11–12:1–30.

Tschermak, E., 1900: "Über künstliche Kreuzung bei *Pisum sativum*," *Ber. deutsche bot. Ges.*, **18**:232–239 (English translation in *Genetics* **35**(1950), Suppl. to No. 5, Part 2:42–47).

————, 1951: "Historischer Rückblick auf die Wiederentdeckung der Mendelschen Regeln," *Verh. zool.-bot. Ges. Wien*, **92**:25–35 (English translation in *J. Hered.*, **42**:163–171).

————, 1956: "Gregor Mendels Versuchstätigkeit und die Zeit der Wiederentdeckung seiner Vererbungsgesetz," *Novant' Anni delle Leggi Mendeliane*, L. Gedda, Ed., Instituto "Gregorio Mendel," Rome.

————, 1960: "60 Jahre Mendelismus," *Verh. zool.-bot. Ges. Wien*, **100**:14–25.

de Vries, Hugo, 1900a: "Sur la fecondation hybride de l'endosperme chez le mais," *Rev. gen. de Bot.*, **12**:129–136.

————, 1900b: "Sur les Unités des caractères spécifiques," *Rev. gen. de Bot.*, **12**:257–269.

————, 1900c: "Sur la loi de disjonction des hybrides," *C. R. de l'Acad. Sci. Paris*, **130**:845–847.

————, 1900d: "Das Spaltungsgesetz der Bastarde (Vorläufige Mitteilung)," *Ber. deutsch. bot. Ges.*, **18**:83–90.

————, 1901–03: *Die Mutationstheorie. Versuche und Beobachtungen über die Entstehung der Arten im Pflanzenreich*, I Bd. *Die Entstehung der Arten durch Mutation*. II Bd. *Elementare Bastardlehre*, Veit u. Co., Leipzig. (English translation by J. B. Farmer and A. D. Darbishire, *The Mutation Theory: Experiments and Observations on the Origin of Species in the Vegetable Kingdom*, Open Court, Chicago, 1909–10.)

————, 1910; *Intracellular Pangenesis*, Open Court, Chicago. Translation from the German edition of 1889 by C. S. Gager, including a translation of "Befrüchtung und Bastardierung" (1903).

————, 1918–1927: *Opera e periodicis collata*, 7 vols., A Oosthoeck, Utrecht. (Contains texts of all of de Vries's papers as published in journals.)

Walker, J. C., 1951: "Genetics and Plant Pathology," in L. C. Dunn,

Ed., *Genetics in the 20th Century,* The Macmillan Company, New York.

Wallace, Alfred Russel, 1889: *Darwinism,* The Macmillan Company New York.

Watson, J. D., and F. H. C. Crick, 1953: "The Structure of DNA," *Cold Spring Harbor Symp. Quant. Biol.,* **23**:123–131.

Weinberg, Wilhelm, 1908: "Über dem Nachweis der Vererbung beim Menschen," Jahreshelfte *Ver. vaterl. Naturk. Württemberg,* **64**:369–382. (English translation in Boyer 1963).

Weinstein, Alexander, 1958: "Did Naegeli Fail to Understand Mendel's Work?" *Proc. X Int. Congr. Genetics,* **2**:339.

———, 1959: "The Geometry and Mechanics of Crossing Over," (with an appendix: Naegeli's rejection of Mendel's theory), *Cold Spring Harbor Symposia,* **23**:177–196.

———, 1962: "The Reception of Mendel's Paper by His Contemporaries," *Proc. 10th Cong. Hist. Sci.,* pp. 997–1001.

Weismann, August, 1883: *Über die Vererbung,* G. Fischer, Jena.

———, 1893: *The Germ-Plasm, a Theory of Heredity,* Trans. by W. N. Parker and H. Rönnfeldt, Walter Scott, Ltd., London.

———, 1904: *The Evolution Theory,* English translation by Prof. J. A. Thomson and Margaret R. Thomson, 2 vols., London.

Weldon, W. F. R., 1902a: "Mendel's Laws of Alternative Inheritance in Peas," *Biometrika,* **1**:228–254.

———, 1902b: "On the Ambiguity of Mendel's Categories," *Biometrika,* **2**:44–55.

Wilson, E. B., 1900: *The Cell in Development and Inheritance,* 2nd ed., The Macmillan Company, New York.

———, 1928: *The Cell in Development and Heredity,* 3rd ed., with corrections, The Macmillan Company, New York.

———, and T. H. Morgan, 1920: "Chiasmatype and Crossing Over," *Amer. Nat.,* **54**:193–219 (Part I by E. B. Wilson: "A Cytological View of the Chiasmatype Theory").

Winge, O., 1958: "Wilhelm Johannsen," *J. Hered.,* **49**:83–88.

Woltereck, R., 1909: "Weitere experimentelle Untersuchungen über Artveränderung, speziell über das Wesen quantitativer Artunterschiede bei Daphniden," *Verh. deutsch. zool. Ges.,* **19**:110–173.

———, 1928: "Bemerkungen über die begriffe 'Reaktionsnorm' und 'Klon'," *Biol. Zentralbl.,* **48**:167–172.

Wright, Sewall, 1917: "Color Inheritance in Mammals. VI Cattle," *J. Hered.,* **8**:521–527.

———, 1918: "Color Inheritance in Mammals. XI Man," *J. Hered.,* **9**:227–240.

———, 1921: "Systems of Mating I–V," *Genetics,* **6**:111–178.

———, 1922: "The Effects of Inbreeding and Cross-Breeding on Guinea-Pigs," *Bull. 1090 U. S. Dept. Agric.* 1–63 and Bull. 1191 ibid. 1–59.

———, 1923: "Mendelian Analysis of the Pure Breeds of Livestock," *J. Hered.,* **14**:339–348 and 405–422.

————, 1931: "Evolution in Mendelian Populations," *Genetics*, 16: 97–159.

————, 1932: "The Roles of Mutation, Inbreeding, Crossbreeding and Selection in Evolution," *Proc. 6th Intern. Cong. Genetics*, 1: 356–366.

————, 1941: "The Physiology of the Gene," *Physiol. Rev.*, 21:487–527.

————, 1951: "Fisher and Ford on 'The Sewall Wright Effect'," *Amer. Scientist*, 39:452–458.

————, 1959a: "Genetics, the Gene and the Hierarchy of Biological Sciences," *Proc. X Int. Cong. Genetics*, 1:475–489.

————, 1959a: "Physiological Genetics, Ecology of Populations and Natural Selection," *Perspectives in Biol. and Med.*, 3:107–151.

————, 1960: "Genetics and Twentieth Century Darwinism," A Review and Discussion, *Amer. J. Hum. Gen.*, 12:365–372.

Yule, G. Udny, 1902: "Mendel's Laws and their Probable Relation to Inter-Racial Heredity," *New Phytologist*, 1:193–207, 222–238.

————, 1906: "On the Theory of Inheritance of Quantitative Compound Characters on the Basis of Mendel's Laws—A Preliminary Note," *Report 3rd Int. Conf. Genetics.**

Zirkle, Conway, 1935: *The Beginnings of Plant Hybridization*, University of Pennsylvania Press, Philadelphia.

————, 1946: "The Early History of the Idea of the Inheritance of Acquired Characters and of Pangenesis," *Transact. Amer. Philos. Soc.*, New Series 35:91–151.

————, 1951a: "The Knowledge of Heredity before 1900," in Dunn, L. C., Ed., *Genetics in the 20th Century*, The Macmillan Company, New York.

————, 1951b: "Gregor Mendel and His Precursors," *Isis*, 42:97–104.

————, 1964: "Some Oddities in the Delayed Discovery of Mendelism," *J. Hered.*, 55:65–72.

— Index —